Christian Reinsch

Amphoteric Liposomes for the Delivery of Oligonucleotides

Christian Reinsch

Amphoteric Liposomes for the Delivery of Oligonucleotides

Elementary studies in vitro and in vivo

Südwestdeutscher Verlag für Hochschulschriften

Impressum / Imprint
Bibliografische Information der Deutschen Nationalbibliothek: Die Deutsche Nationalbibliothek verzeichnet diese Publikation in der Deutschen Nationalbibliografie; detaillierte bibliografische Daten sind im Internet über http://dnb.d-nb.de abrufbar.
Alle in diesem Buch genannten Marken und Produktnamen unterliegen warenzeichen-, marken- oder patentrechtlichem Schutz bzw. sind Warenzeichen oder eingetragene Warenzeichen der jeweiligen Inhaber. Die Wiedergabe von Marken, Produktnamen, Gebrauchsnamen, Handelsnamen, Warenbezeichnungen u.s.w. in diesem Werk berechtigt auch ohne besondere Kennzeichnung nicht zu der Annahme, dass solche Namen im Sinne der Warenzeichen- und Markenschutzgesetzgebung als frei zu betrachten wären und daher von jedermann benutzt werden dürften.

Bibliographic information published by the Deutsche Nationalbibliothek: The Deutsche Nationalbibliothek lists this publication in the Deutsche Nationalbibliografie; detailed bibliographic data are available in the Internet at http://dnb.d-nb.de.
Any brand names and product names mentioned in this book are subject to trademark, brand or patent protection and are trademarks or registered trademarks of their respective holders. The use of brand names, product names, common names, trade names, product descriptions etc. even without a particular marking in this works is in no way to be construed to mean that such names may be regarded as unrestricted in respect of trademark and brand protection legislation and could thus be used by anyone.

Coverbild / Cover image: www.ingimage.com

Verlag / Publisher:
Südwestdeutscher Verlag für Hochschulschriften
ist ein Imprint der / is a trademark of
AV Akademikerverlag GmbH & Co. KG
Heinrich-Böcking-Str. 6-8, 66121 Saarbrücken, Deutschland / Germany
Email: info@svh-verlag.de

Herstellung: siehe letzte Seite /
Printed at: see last page
ISBN: 978-3-8381-3608-0

Zugl. / Approved by: Halle, MLU, Diss., 2011

Copyright © 2013 AV Akademikerverlag GmbH & Co. KG
Alle Rechte vorbehalten. / All rights reserved. Saarbrücken 2013

to my companions
Una, Silke & Andreas

*"Science is not a creed.
It was not revealed to man by some superior deity.
Science is a product of the human brain, and as such,
it is always open to discussion and possible revision.
(…) We select experimental results that appear to us
as logically connected together, and we ignore many
facts that do not fit into our "logic". This rather artificial
procedure is our own invention and we are so proud of
it that we insist its results should be considered as
"laws of nature"."*

Leon Brillouin
Scientific Uncertainty, and Information (1964)

TABLE OF CONTENTS

1	**Introduction**	**7**
1.1	Challenges associated with harnessing oligonucleotides	7
1.2	Antisense oligonucleotides (ASO)	7
1.3	RNA interference (RNAi)	10
1.4	Unassisted application and membrane crossing of oligonucleotides	12
1.5	Lipid-based vehicles for oligonucleotide delivery	13
1.5.1	Lipid-based vehicles (Liposomes) – an overview	14
1.5.2	Amphoteric Liposomes (Smarticles)	15
1.5.3	Principal considerations on the PK and BD of liposomes	16
1.6	Lipid shape theory & mechanism of pH-sensitive liposomes	18
1.7	ApoB100 – a valuable target to investigate oligonucleotide-mediated gene silencing *in vivo*	21
1.8	Scope of the thesis	22
2	**Materials & Methods**	**23**
2.1	Materials	23
2.1.1	Lipids	23
2.1.2	Fluorescence dyes	24
2.1.3	Oligonucleotides	25
2.1.4	Antibodies	26
2.1.5	Cells and cell culture	26
2.2	Methods	27
2.2.1	Preparation of liposomes	27
2.2.1.1	Alcohol injection	27
2.2.1.2	Concentration and separation	27
2.2.2	Characterization of liposomes	29
2.2.2.1	Particle size determination	29
2.2.2.2	Determination of zeta potential	29
2.2.2.3	Determination of lipid concentration	29
	A) PHOSPHATE-Test	
	B) CHOL-CHOD-PAP-Test	
2.2.3	Determination of oligonucleotide concentrations	29
2.2.4	Determination of non-encapsulated oligonucleotides	30
2.2.5	*In vitro* studies using primary mouse hepatocytes	31
2.2.5.1	Isolation of primary mouse hepatocytes	31
2.2.5.2	*In vitro* transfection of primary mouse hepatocytes	31

2.2.6	Animal trials	32
2.2.6.1	Pharmacokinetic (PK) and Biodistribution (BD) study	32
2.2.6.2	Pharmacodynamic study in mouse liver ("nov038-LT1-ASO")	33
2.2.6.3	Pharmacodynamic studies in mouse liver and plasma (ApoB100 trials)	33
2.2.7	Total body and organ scan	34
2.2.8	Determination of Cy5.5 fluorescence signals in blood samples (PK)	34
2.2.9	Determination of Cy5.5. fluorescence signals in tissue samples (BD)	35
2.2.10	Cryosections	35
2.2.11	Fluorescence microscopy	36
2.2.12	Confocal laser scanning microscopy (CLSM)	36
2.2.13	Determination of plasma values (liver enzymes and plasma cholesterol)	36
2.2.14	Cytokine ELISA	37
2.2.15	Quantification of mRNA	37
2.2.15.1	QuantiGene (QG)	37
2.2.15.2	Real-time PCR	39
2.2.16	Western Blot analysis	39
2.2.16.1	Sample preparation for protein analysis	39
	A) LT1 protein	
	B) ApoB100 protein	
2.2.16.2	Western Blot analysis	40
2.2.17	Statistical analyses	41
3	**Results**	**42**
3.1	Pharmacokinetic and Biodistribution of nov038	42
3.1.1	Whole body imaging indicates a fast distribution into liver and spleen	43
3.1.2	Pharmacokinetic of free and encapsulated Cy5.5-labeled ASO	45
3.1.3	Quantitative organ distribution	48
3.1.4	Microscopic distribution	49
3.1.5	Determination of plasma AST/ALT levels and proinflammatory cytokines	51
3.2	Pharmacodynamic of nov038-LT1-ASO	53
3.3	Proof-of-concept study using nov038-ApoB-siRNA	57
3.4	*In vitro* transfection of primary mouse hepatocytes (PMHs) using nov038 loaded with either ASO or siRNA	59
3.4.1	Transfection of PMHs with ASO and siRNA molecules targeting apoB100 mRNA using the cationic transfectant jetPEI™-Gal	59
3.4.2	Transfection of PMHs with ASO and siRNA molecules targeting apoB100 mRNA encapsulated into nov038	60
3.4.3	Uptake of nov038 loaded with Cy5.5-labeled ASO or siRNA by PMHs	62

3.5	Rational design of Smarticles and application for PMHs	64
3.6	Delivery of oligonucleotides using the fusogenic nov729	67
3.6.1	Transfection of PMHs with nov729 encapsulating ASOs or siRNAs	67
3.6.2	Pharmacodynamic of nov729 laoded with ApoB I 5'P siRNA *in vivo*	69
3.6.3	Transfection of PMHs with nov729 in the presence of mouse serum	72
4	**Discussion**	**74**
4.1	PK of nov038 is non-linear and depends on lipid dose	74
4.2	Nov038 distributes into saturable compartments	75
4.3	Free, non-encapsulated (naked) ASO shows a rapid kinetic	76
4.4	Microscopy reveals uptake of nov038 by the liver parenchyma	77
4.5	High lipid doses of nov038 are non-toxic	78
4.6	Nov038 delivers ASO but not siRNA molecules	79
4.7	A rational design of fusogenic liposomes enables the effective delivery of siRNAs on PMHs	81
4.8	Delivery of siRNA *in vivo* by using nov729 is inefficient and inhibitable by mouse serum	83
4.9	Conclusions and future perspectives	85
5	**Summary**	**87**
6	**References**	**88**
7	**Abbreviations**	**98**
8	**Appendix**	**100**
9	**Acknowledgement**	**103**

1. Introduction

1.1 Challenges associated with harnessing oligonucleotides

Oligonucleotides represent a class of biomacromolecules being effective by a new pharmacology. Especially single-stranded antisense oligonucleotides (ASO) or double-stranded small interfering RNAs (siRNA) bind to cognate RNA sequences through Watson-Crick hybridization resulting in the inhibition of the protein-coding target RNA. Thereby, ASOs and siRNAs activate the targeted, enzyme-mediated degradation of the mRNA and belong to the category of so-called "informational drugs" in which the drug specificity is coded by the sequence and not by its molecular structure. Never before has the "receptor" mRNA been considered in the context of drug receptor interactions. In contrast to small-molecule drugs or antibodies which directly interfere with a disease-mediating protein, siRNAs and antisense drugs impede the *de-novo* synthesis of proteins and thus act at an earlier phase in the disease-fighting process. Whereas small-molecules and antibody drugs are usually restricted to extracellular targets, oligonucleotides can be designed to interfere with every (therapeutic interesting) gene or mRNA. After decoding the human genome the design of oligonucleotides also succeeded in being more rapid, less complex and more efficient than traditional drug design targeting proteins.

However, rapid nucleolytic degradation of oligonucleotides in body fluids and a high hydrophilicity which hinders them from effective membrane crossing often requires the assisted transport of oligonucleotides by delivery systems such as liposomes. They specifically address challenges involved with the transit of oligonucleotides, namely biodistribution, cellular uptake and endosomal release, which is also subject of the present work.

1.2 Antisense oligonucleotides (ASO)

ASOs, being usually 15 to 20 (desoxy)-ribonucleotides in length, specifically inhibit gene expression by Watson-Crick base pairing to their complementary (pre-)messenger RNA. As illustrated in Fig. 1.1, two major mechanisms of action have been elucidated performing post-transcriptional gene silencing: A) After binding to the (pre-) mRNA most of the single-stranded ASOs are designed to mediate the cleavage of the DNA:RNA hybrid by RNAse H1. The endo-ribonuclease is primarily located in the nucleus and cleaves the RNA moiety of this heteroduplex with subsequent degradation of the target mRNA.[1] B) Those ASOs which do not induce the RNase H1 cleavage were customized to inhibit the translation of the mRNA by a steric blockade of the ribosome. Most of these ASOs were directed against the 5'-terminus (cap region) or the AUG initiating codon region of the target mRNA and prevented the binding and assembly of the translation machinery very efficiently.[2]

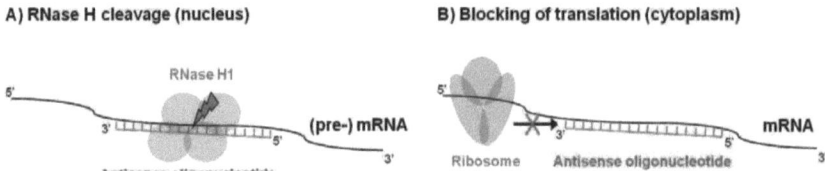

Fig. 1.1: **Mechanisms of antisense activity.** A) RNase H1 cleavage of the target mRNA induced by ASO molecules. B) Inhibition of translation by steric blocking of the ribosome. (modified from [3])

Besides the identification of accessible target sites at the mRNA the type and degree of chemical modification of the antisense molecule affect its mode of action (Tab. 1.1). Generally, three types of modifications can be distinguished: an altered phosphate backbone, various sugar modifications (especially at the 2' position of the ribose), and unnatural base-analogs. The use of base-modified antisense technologies was reviewed by Herdewijn [4] and the following section will focus on oligonucleotides with modified sugar moieties and phosphate backbones.

Phosphorothioate (PS) oligonucleotides, in which one of the non-bridging oxygen atoms on the phosphodiester bond is replaced by sulfur, are the best known and widely used representatives of the 'first generation' of antisense molecules (Tab. 1.1). The introduction of PS linkages into the DNA backbone improved resistance of the ASO towards nucleases and increased the half-life ($t_{1/2}$) in human serum by about 10-fold.[5,6] The backbone sulfur is accessible for plasma proteins and mediates the binding to α2-macroglobulin and albumin.[7] Once hooked onto these natural and abundant carriers, PS modified oligonucleotides escape from rapid renal excretion and accumulate mainly in the cortex and medulla of the kidneys, in the liver, lymph nodes and spleen.[8] Only little material migrates into lung, colon and ileum.[9] PS oligonucleotides still bind to target mRNA and mediate cleavage by RNAse H1.[10]

While the phosphorothiolation solved the most pressing issues, a number of drawbacks remained or newly appeared: (1) After intravenous (iv) bolus injections PS oligonucleotides mediate unspecific protein interactions (causing complement activation, hypotension, thrombocytopenia), largely because of their polyanionic backbone and (2) they exhibit a reduced mRNA binding affinity when compared to their unmodified counterparts.[11-13] (3) Further, PS oligonucleotides are still extensively digested in plasma and tissues by exonucleases.[14,15]

	1st Generation	2nd Generation		3rd Generation
	Phosphorothioate DNA (PS)	2'-O-methyl RNA (2'OMe)	2'-O-methoxy-ethyl RNA (2'MOE)	Locked nucleic acid (LNA)
serum protein binding	++	+/-	+/-	+/-
nuclease resistance	+	++	++	++
affinity to mRNA	-	++	++	+++
activation of RNase H1	+	-	-	-
References	5-7,10,16	17,18	19-21	22,23

Tab. 1.1: Generations of antisense modifications. Further details are noted in the text. B indicates one of the bases adenine, guanine, cytosine or thymine. Compared to the unmodified phosphodiester (PO) DNA molecule $^{+/++/+++}$ denotes the degree of improvement, $^{+/-}$ no changes and $^-$ indicates an impairment of the respective modification.

A second generation of PS ASOs with modifications at the 2' position of the ribose such as 2'-O-methyl (2'OMe) or 2'-O-methoxy-ethyl (2'MOE) exhibits higher affinity towards the complementary RNA and higher nuclease resistance.[17,18] For example 1st generation ASO molecules show a tissue half-life of only 1-2 d [24] whereas 2nd generation ASOs (e.g. 2'MOE modified) exhibit a longevity of 8-22 d in target tissues, depending on dose and tissue type.[19,21,25] Sugar modifications such as 2'OMe or 2'MOE are not compatible with RNAse H1 activity and therefore need to be restricted to the wings of the oligonucleotide molecule leaving a central window of 2' unmodified DNA nucleotides.[26] The so-called gapmers, chimeric DNA-MOE oligonucleotides, with at least five nucleotides between the modified wings (e.g. 20mers with 5^{MOE}-10^{DNA}-5^{MOE}) increase the ASO potency by 5-15 fold compared to its phosphorothiolated counterpart *in vitro* and *in vivo* and represent the current state-of-the-art ASOs for clinical use.[27]

Locked nucleic acids (LNA) represent a novel class of nucleic acid analogues subsumed under the term 'third generation' of antisense agents. The "lock" is a methylene bridge connecting the 2'-oxygen with the 4'-carbon of the ribose molecule.[28] Introduction of LNA into a DNA oligonucleotide induces a conformational change of the DNA:RNA duplex towards the A-type helix and therefore prevents RNase H1 cleavage of the target RNA.[29] Like the abovementioned 2'MOE modified ASO, LNA gapmers exhibit increased stability against nucleases and unprecedented binding affinity towards complementary DNA or RNA.[22] This improves RNAse H1 cleavage and leads to a higher potency of LNA gapmers in gene silencing compared to the 2'MOE gapmers.

1.3 RNA interference (RNAi)

RNA interference, a natural occurring phenomenon, is an evolutionary conserved mechanism for post-transcriptional gene silencing (Fig. 1.2). It was first described in the nematode *Caenorhabditis elegans* in the late 90's [30] and has been demonstrated in diverse eukaryotes such as insects, plants, fungi and vertebrates.[31] Fire & Mellow injected long double-stranded RNA (dsRNA) into the gonads of *C. elegans* to initiate RNA interference [30] and Tuschl and coworkers demonstrated that small interfering RNAs (siRNAs), processed into 21-23 nucleotides long RNAs, can specifically suppress gene expression in mammalian cells.[32] Within the cellular RNAi pathway long dsRNA is cleaved into smaller fragments of 20-30 nucleotides with two-nucleotide 3'- or 5'-overhanging ends by the highly conserved endonuclease Dicer, located in the cytoplasm.[33,34] The short dsRNAs (e.g. siRNAs or microRNAs [miRNAs]) generated that way are subsequently incorporated into the RNA-induced silencing complex (RISC), a multi-functional protein:RNA complex.[35] Active RISC complexes (RISC*) promote the unwinding of the siRNA through an ATP-dependent process and the unwound antisense strand guides RISC* to the complementary mRNA.[36] The mRNA of the antisense:sense duplex is than cleaved through hydrolysis at a single site by the nuclease Argonaute, the core constituent of the RISC.[37] Finally, the cut mRNA is degraded by intracellular RNAses and is not available for further translation processes.

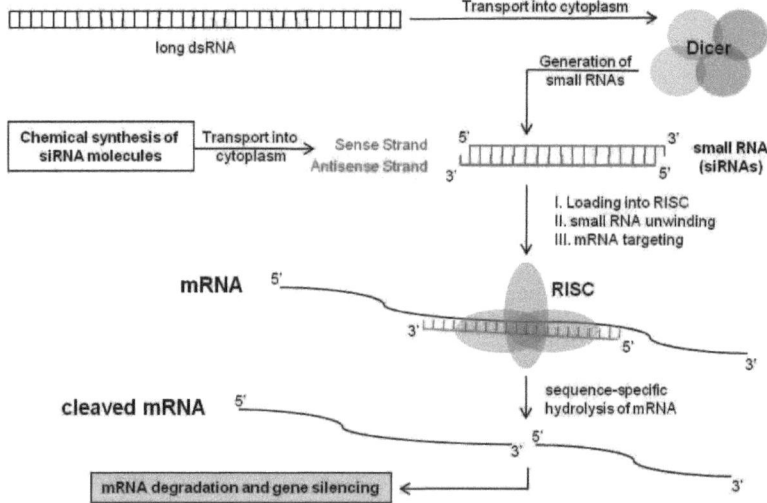

Fig. 1.2: Gene silencing by siRNA molecules. RNA interference is induced by siRNAs processed from long dsRNAs or directly delivered into the cytoplasm from an exogenous source.

Originally, siRNAs are processed from dsRNA precursors (from exogenous sources, e.g. long dsRNA, plasmids) and, further, synthetic siRNA molecules can be directly delivered into the cells. Importantly, RNAi offers a higher potency than antisense strategies as the effector molecules such as siRNAs may function at much lower concentration and the effect is long lasting. Once engaged in the RISC, siRNAs can last for weeks which may benefit therapeutic approaches.[38,39]

The elucidation of the RNAi pathway cleared the way for the scientific community to use RNAi as a research tool to temporarily suppress the gene of interest. The breakthrough that RNAi works in mammalian cells [32] led to intense investigation into its role in mammalian cell function, its use as a valuable "lab tool" in cells and animals, and its application for therapeutic purposes.[40] However, RNAi effector molecules, esp. siRNAs, have to overcome numerous hurdles and barriers within the extra- and intracellular environment:

1) Unmodified siRNAs exhibit a very short half-life *in vivo* (~ minutes) due to degradation by serum RNAse A-like enzymes [41] and renal elimination.[42] Lessons were learned from ASO drug development and, for example, the insertion of phosphorothioates into the siRNA backbone prolongs the serum half-life and improves the pharmacokinetic at all.[42] Further, alternating 2'OMe modifications on both strands [43] or the incorporation of several LNAs [44] lead to significant resistance against serum-derived nucleases without the loss of RNAi activity.

2) The unintended ("off-target") silencing of numerous transcripts which share partial complementary to the siRNA duplex is a widespread phenomenon and limits the specificity of siRNAs for functional genomics and therapeutic applications.[45] A stringent sequence selection and a smart modification (e.g. 2'OMe or LNA) favors the incorporation of the antisense strand into RISC and diminishes the risk of off-target effects.[46,47]

3) Additionally, siRNA molecules can trigger cells of the immune system to produce cytokines both *in vitro* and *in vivo* causing undesirable side effects.[48] Recent findings propose the involvement of toll-like receptors (TLR), located within the endosome of mammalian immune cells, during sensing of foreign DNA and RNA.[49,50] Whereas dsRNA is recognized by TLR3 in a sequence-independent manner, TLR7 and TLR8 perceive ssRNA and dsRNAs as short as 19-21 nucleotides (siRNAs) in a sequence-specific way.[48,51] Being part of the innate immunity the signaling via this subfamily of TLRs upon recognition of nucleic acids proceeds through intracellular pathways [52] leading to the induction of proinflammatory cytokines (tumor necrosis factor alpha (TNF-α), interleukin (II) 6, Il-12) and interferon alpha (IFN-α).[51,53] Here, Judge and co-workers provide a basis for the rational design of siRNAs avoiding the activation of the innate immune response.[51,54]

4) Finally, getting siRNAs into the cytoplasm is the most challenging hurdle as they are 10-30 times greater than typical small molecule drugs and highly charged and thus too hydrophilic and bulky to overcome the cell membrane by nature.

1.4 Unassisted application and membrane crossing of oligonucleotides

The uptake of ASOs into cells is not well defined yet but there are some evidences that plasma proteins hooked with ASOs interact with cell surface proteins including serum albumin with megalin (LRP2) and α2-macroglobulin with the low-density lipoprotein receptor-related protein 1 (LRP1) that enhances crossing of the plasma membrane.[2,55] Once bound to a cell membrane ASOs are internalized by (receptor-mediated) endocytosis and reach either endosomal or lysosomal vesicles.[55] The escape mechanisms from the vesicular pathways are not fully understood but it is an essential prerequisite of oligonucleotides to run off from the endosome and/or lysosome intact to exert sequence-specific antisense effects within the cytoplasm or nucleus. Therefore, Akhtar and co-workers already speculated in 1991 that the efflux of oligonucleotides from the endosome is mediated by one or more proteins present in the bilayers.[56] In 2010, Bennett and Swayze proposed a protein complex or channel, called "oligoportin", which allows the passage of ASOs.[2]

Stein and co-workers recently published a novel method for the "naked" delivery of LNA-modified oligonucleotides *in vitro* and *in vivo* (Tab.1.2), called "gymnosis" (from *gymnos* [greek] = naked). To promote efficient ASO uptake cells were seeded and transfected at low plating density and high ASO concentrations (2.5 – 10 µM) were used.[57] However, a molecular mechanism for ASO uptake was not revealed.

Numerous preclinical trials allocate the pharmacological activity of unassisted ASOs following systemic or topic application. However, ASO-mediated gene silencing is only effective at high dosages and usually chemically stabilized ASOs of the 2nd or 3rd generation are used. In rodents, dosages for 1st generation ASO ranged from 10 to 75 mg/kg/day, whereas 2'MOE- or LNA-modified oligonucleotides are injected at dosages from 5 to 50 mg/kg/week, depending on the target tissue (see also Tab. 1.2).[20,58] Consistently, clinical phase III studies using a 2nd generation chimeric ASO against apoB100 (Mipomersen®, ISIS301012) demonstrated a dose-dependent reduction in plasma ApoB100 levels with dosages of 50 to 400 mg per week in humans.[59,60]

It is widely accepted that naked, unmodified siRNA is unable to passively cross the cell membrane and is thus far not active *in vivo* after systemic injection. High pressure or hydrodynamic (tail vein) injections of naked siRNAs elicit target gene knockdown especially in the liver of mammalian model organisms.[61,62] However, the hydrodynamic intravenous injection requires large volumes (~1 ml / mouse) administered at high pressure over a short period of time and is thus not applicable for human application.

In 2002, Hunter and co-workers introduced the multispan transmembrane protein SID-1 (*s*ystemic RNA *i*nterference-*d*eficient) as a putative cell membrane transporter of double-stranded RNA in *C. elegans*.[63,64] SID-1 is required for the systemic RNAi; it passively transports dsRNA into the cytoplasm like a pore or channel [63] independent of chain length [65] and is therefore supposed to act as siRNA transporter. The mammalian homolog of SID-1 (FLJ20174) has been demonstrated to enhance the uptake of siRNA molecules in human cancer cells [66] and silencing of SID-1 in human hepatic cells reduced the internalization of lipophilic siRNAs.[67] However, the cellular uptake of naked (modified) siRNA remains unproven and publications describing the transporter SID-1 are restrained.

To this date systemically injected naked siRNA molecules failed to yield endpoints in animal models and studies were focused on local / topical siRNA delivery. Intravitreally injected siRNAs targeting the VEGF pathway in wet age-related macular degeneration (AMD), a retinal disease causing loss of vision, were under clinical investigation.[68] However, the trials of both candidates (Bevasiranib@ (Cand5) by OPKO Health, Inc. and Sirna-027 (AGN211745) by Allergan, Inc.) were terminated after poor phase II/III data.

In summary, the effectiveness of 2nd generation antisense molecules either after systemic or local injections is widely accepted and enables convenient application at low dosages in selected tissues (e.g. liver). Further, several 2'-modified chimeric anti-cancer ASO molecules are currently under clinical investigation.[2] However, the biodistribution of ASO molecules far from liver, spleen and kidney is poor and applications in cancer or inflammation diseases require high dosages.[13,69]

The challenge of siRNA uptake requires the assisted transport through the body. Nucleases rapidly degrade siRNAs in biological fluids; they are quickly excreted via the kidneys and induce immune responses after recognition by endosomal TLRs. Further, siRNA molecules are not able to cross the plasma membrane by themselves. Formulation of siRNAs with proper delivery systems can solve most of these pressing issues and can further direct them to the appropriate tissues. The next chapters focus on strategies for the delivery of ASO and siRNA molecules, namely by liposomes.

1.5 Lipid-based vehicles for oligonucleotide delivery

Significant progress has been made in the construction of delivery systems that enable cytosolic delivery or nuclear uptake of oligonucleotides without affecting cellular integrity. Reinsch et al., 2008, Wu & McMillian, 2009 as well as Reischl & Zimmer, 2009 profile some of the most advanced non-viral delivery vehicles for oligonucleotides including synthetic oligonucleotide conjugates, polymer- or lipid-based systems.[70-72] The following chapter focus on the delivery of oligonucleotides by using lipid-based vehicles and highlights common building and mechanistic principles.

1.5.1 Lipid-based vehicles (Liposomes) – an overview

Liposomes, characterized by a bilayered membrane assembly, are mainly made of phospholipids bearing a diacyl-glycerol membrane anchor or cholesterol derivatives. Within the last 30 years liposomes have been developed as a pharmaceutical carrier for therapeutic agents including small molecules, proteins and DNA/RNA-based drugs and several FDA-approved liposomal formulations are presently available on the market (AmBisome®, Doxil/Caelyx®, Visudyne® and others).

The most common feature of all oligonucleotide carriers, either lipid- or polymer-based, is a positive surface charge, which facilitates rapid complex formation with negatively charged oligonucleotides resulting in high weight ratios between cargo and vector. In addition, complexes with a cationic net-charge are readily adsorbed onto cells, leading to a high local oligonucleotide concentration at the cell surface, which supports internalization.

Stable nucleic acid-lipid particles, SNALPs, are PEGylated cationic lipid carriers originally comprising the ionizable 1,2-dilinoleyloxy-3-dimethylaminopropane (DLinDMA) specially developed by Protiva Biotherapeutics Inc. (now *Tekmira Pharmaceuticals Corp.*, Burnaby, BC, CA). Recent developments created the ionizable cationic DLin-KC2-DMA which was formulated in SNALPs and showed to have *in vivo* activity at siRNA doses as low as 0.01 mg/kg in rodents and 0.1 mg/kg in nonhuman primates.[73] SNALPs PEG-lipids with rather short membrane anchors exhibit sufficient membrane residence during production and storage, but redistribute in the presence of a sink such as lipoproteins or cellular membranes.[74]

Alnylam Pharmaceuticals, Inc. (Cambridge, MA, USA) has been developed a combinatorial library comprising lipid-like agents varying in i) alkyl chain length, ii) ester or amide linkages between the alkyl chains and the amine and iii) the polar amine-containing head group were tested *in vitro* and in rodents and nonhuman primates. For *in vivo* testing of nanoparticles lipidoid materials were formulated with cholesterol, PEG-lipids and 2'OMe modified siRNAs targeting coagulation Factor VII and ApoB100 [75,76] or PCSK9 mRNA [77] for preferably liver delivery after intravenous injection (see also Tab. 1.2). A recent publication reported therapeutic efficacy of epoxide-derived lipidoids with pyrazine containing amine head groups (called C12-200) at dosages of less than 0.1 mg/kg in mouse hepatocytes indicating a hundredfold improvement in potency over the prior lipidoids.[78]

In contrast, the delivery profile of neutral liposomes consisting of 100 % dioleoyl-phosphatidylcholine (DOPC) [79,80] or a mixture of egg phosphatidylcholine (PC) [81] and cholesterol was investigated in mouse models of cancer or inflammation. Since these vectors lack the electrostatic interaction between cargo and carrier, efficient sequestration of the oligonucleotides during production and strict confinement after injection are typical challenges in this group of vectors.

1.5.2 Amphoteric liposomes (Smarticles)

Smarticles (formerly proprietary of novosom AG, now Marina Biotech Inc., Bothell, WA, USA) are charge-reversible (amphoteric) lipid-based formulations and respond to the pH of the environment. Being negatively charged at neutral pH, amphoteric liposomes share the biodistribution properties of known anionic or neutral liposomes. However, once taken up by a cell and exposed to low pH in the endosome, amphoteric liposomes become neutral and eventually cationic and thus provide a mechanism for endosome release and intracellular delivery of sequestered oligonucleotides.[82-84] Further, the cationic surface charge at pH 4-5 is used for efficient loading of oligonucleotides during the encapsulation process (called "advanced loading procedure", ALP).

The lipid mixture of Smarticles formulations comprises distinct portions of anionic and cationic lipids either amphoteric or permanently charged and neutral lipids resulting in three classes of amphoteric liposomes:

Amphoter I: comprising a permanent cationic lipid and a charge-reversible anionic lipid
Amphoter II: comprising both charge-reversible anionic and cationic lipids
Amphoter III: comprising permanent anionic lipid and a charge-reversible cationic lipid

Safe and efficient *in vivo* delivery has been demonstrated in mouse models of colitis[85] and collagen-induced arthritis [86] using CD40 targeting antisense oligonucleotides encapsulated into Smarticles formulation nov038. This Amphoter II class liposome is based on the fully charge-reversible lipids α-(3-O-cholesteryloxy)-δ-(N-ethylmorpholine)-succinamide (MoChol, cationic) in combination with the cholesteryl-hemisuccinate (CHEMS, anionic) and neutral lipids dioleoyl-phosphatidylethanolamine (DOPE) and palmitoyl-oleoyl-phosphatidylcholine (POPC) at molar ratios of 20:20:45:15. Equimolar mixtures of the charged lipids were found to stably sequester oligonucleotides and a 3:1 combination of DOPE to POPC substantially improved the serum stability.[86]

Systemic administration of nov038 effectively delivers ASO to the liver, spleen and sites of inflammation [87] and treats established arthritic disease by improving clinical parameters, inflammation and joint damage. The therapeutic efficacy of nov038-CD40-ASO is related to its tropism for monocytes/macrophages and myeloid dendritic cells, where it results in rapid down-regulation of CD40, reduction of major inflammatory cytokines such as TNFα, IL-6 and IL-17 and inhibition of T cell responses in draining lymph nodes.[86] Further, nov038 mediated the delivery of ASO molecules to hepatocytes and potentiated the antisense dependent knockdown of several target genes (this work & unpublished data) compared to non-encapsulated ASO molecules.

1.5.3 Principal considerations on the pharmacokinetics (PK) and biodistribution (BD) of liposomes

A blood borne drug has to be able to leave the vasculature in order to be distributed inside the target tissue. Liposomes have been widely used to alter the pharmacokinetic and biodistribution profile of encapsulated drugs in circulation. Factors including charge, size, dose and lipid composition are well known parameters influencing that carrier profile.[88-93] Both, cationic and anionic particles are removed from the bloodstream by cells of the mononuclear phagocyte system (MPS, formerly known as reticuloendothelial system, RES) located in the liver (Kupffer cells) and in the spleen.[90,94] Cationic liposomes, for example, tend to form large aggregates with anionic serum components (e.g. complement proteins) that are cleared rapidly from the circulation or associate directly with the glycoprotein layer of the endothelium and get trapped in first pass organs.[95,96] Major organs of distribution after iv administration of cationic liposomes are the liver and lungs followed by spleen, kidney and heart.[97,98] The accumulation of large cationic aggregates in the lung capillaries where liposomes and cargo are then absorbed can lead to a massive obstruction with fatal consequences. Neutral and negatively charged liposomes distribute mainly in liver and spleen and exhibit longer circulation times but without affecting the lung capillaries or other endothelia. Generally, electrostatically charged liposomes disappear faster from the blood than uncharged liposomes.[99] It could be shown that the clearance of liposomes via the complement system (protein-membrane interactions) depends not only on surface charge.[100,101] Lipid head group and acyl chain composition must also influence liposome-protein interactions. Besides phosphatidylglycerol (PG) and phosphatidylinositol (PI)[100], the incorporation of lipid-conjugated PEG considerably inhibits non-specific interactions with serum proteins and cells and tremendously alters the pharmacological properties of the carriers independent of their surface charge.[91,102-105] PEGylated liposomes offer a substantially increased circulation time but also PEG suppresses the binding to cell membranes and limits the cellular uptake. PEG-lipids with a rather short membrane anchor eventually leave the carrier membrane after intravenous injection and the carrier gradually exposes a more and more cationic surface which improves the affinity to anionic cell surfaces (see SNALPs). However, repeated high-dose administration of PEGylated carriers triggers the host immunogenicity which ends up in faster blood clearance.[106,107]

Particle size also effects the PK and BD of vesicular carriers whereas, generally, small liposomes (<100 nm) are eliminated from the blood more slowly than large liposomes.[88,108] Thereby, the complement activation and uptake of liposomes by cells of the MPS strongly depends on the size whereas the particle recognition and clearance increase with increasing size.[109-111] The complement activation requires the assembly and activation of complement proteins. Devine and co-workers suggested that the more curved surfaces of smaller liposomes cannot achieve the proper geometric configuration for efficient complement activation. Substantial differences were observed in the consumption of complement components at liposomal sizes between 100-200 nm. In addition, depending on size and composition of the liposomes, the parenchymal cells of the liver (hepatocytes) may also play a dominant role in the elimination of liposomes from the blood.[112] Numerous open fenestrations allow the passing of liposomes through the hepatic sinusoidal endothelial lining.[113] In rodents, these fenestrations have a size of 100-200 nm [114] and thus allow small liposomes to gain access to the hepatocytes.[111,115] Generally, small long-circulating nanoparticles readily accumulate at sites of vascular leak including tumor vasculature [116] and inflammation sites relying on the "enhanced permeability and retention (EPR) effect".[117]

Finally, the administered lipid dose plays a crucial role in circulation times and distribution of liposomes following systemic injection. Cullis and co-workers showed that increasing lipid doses (16...1600 µmol/kg BW) of neutral or anionic liposomes leads to prolonged blood $t_{1/2}$ which corresponds to a depletion of blood opsonins and subsequently lowers the probability of MPS-mediated clearance.[118] In addition, early works suggest that increasing lipid doses (5...500 µmol/kg BW) saturate non-specific binding sites of the murine liver and spleen and, as a result of the hepatic and splenic saturation, the liposomal blood levels increase.[88] Further, Chow and co-workers propose that the hepatic uptake of small neutral liposomes in mice involves two parallel pathways in which one is saturable mediated by phagocytic Kupffer cells (blocking lipid dose ~16 µmol/kg BW).[119] The other is a non-saturable, pinocytotic uptake pathway mediated by parenchymal cells, favoring this pathway at high lipid doses. Increasing lipid doses (0.005...159 µmol/kg BW) led to a decrease in relative Kupffer cell uptake and concomitant increase in relative hepatocytes uptake.

Liver and spleen are major sites of liposomal distribution. Thereby, the biodistribution strongly depends on size, surface charge, lipid composition and dose of the liposomes. Charged particles and those with increasing size are cleared rapidly from the bloodstream whereas increasing lipid doses facilitate parenchymal liver uptake and prolonged circulation times of the liposomes. Positive surface charges further facilitate rapid complex formation with negatively charged oligonucleotides. Complexes with a cationic net-charge are readily adsorbed onto cells and are thus internalized more easily. However, aggregate formation with serum components and unspecific adsorption to endothelia can lead to a blockage of the (lung) capillaries. In contrast, amphoteric liposomes are stable in serum (pH 7.4) and distribute in the same manner as true anionic liposomes. In contrast to anionic carriers, amphoteric liposomes exhibit a high payload of oligonucleotides (at pH 4-5). PH-sensitive lipids can provoke an endosomal escape of the drug and a molecular mechanism thereof is given in the next chapter.

1.6 Lipid shape theory & mechanism of pH-sensitive liposomes

According to the shape theory provided by Israelachvili and co-workers the thermodynamically favored aggregate of lipid molecules depends on the ratio between their molecular volume of the hydrocarbon chains (v) and the product of the optimal molecular surface area (a_0) and maximum tail length (l_c) which calculates a shape factor (N_s) or "critical packing parameter" (CPP) [120]: $N_s = v / (a_0 \cdot l_c)$

The surface area a_0 is determined by the volume of the head group, its hydration, charge and hydrogen bonding capabilities whereas the chain volume v is dependent on their thermal motion. Possible CPP values and predicted lipid aggregates are listed in Fig. 1.3.

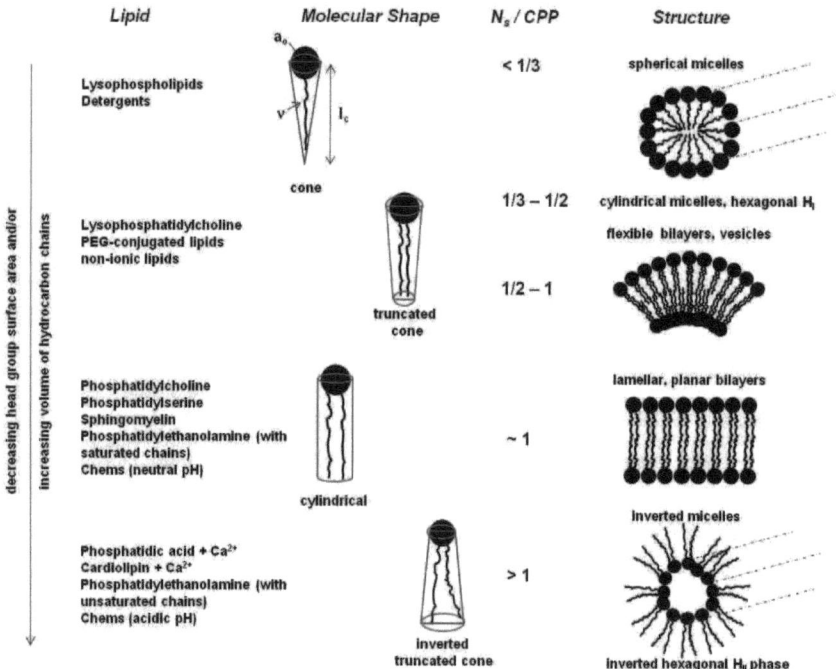

Fig. 1.3: Schematic illustration of lamellar and non-lamellar lipid aggregates formed in aqueous solutions. The "critical packing parameter" CPP defines the molecular lipid shape and their superordinated aggregate structures. (modified from: [120-122])

Liposomes only form when cylindrical molecules predominate or when the tendency of e.g. inverted cone-shaped molecules to form hexagonal structures is balanced by the presence of cone-shaped molecules in the membrane. The non-fusogenic lamellar phase of a lipid bilayer is characterized by a cylindrical shape of lipids whereas the fusogenic hexagonal (H_{II}) phase is characterized by inverted cone-shaped lipids. The Lipid shape theory relates small polar head groups of the lipids to a fusogenic state and larger head groups to a lamellar, non-fusogenic phase.[120,121]

Following encapsulation of drugs and the transport to the target cells liposomes are predominantly internalized by way of endocytosis and end up in the endolysosomal system. For release of the cargo into the cytoplasm the liposomal and endosomal membrane need to fuse, usually in a pH-dependent manner. Thereby liposomes undergo a structural change in order to perform a transition between the stable phase at neutral pH and the fusogenic state at lower pH found within the endosome. Here, the most frequently used concept of membrane-fusion triggered by pH-sensitivity involved the combination of inverted cone-shaped phosphatidylethanolamine (with unsaturated chains, e.g. DOPE) with cylindrical amphiphiles such as CHEMS (see below) which act as a stabilizing agent at neutral pH (reviewed in [123,124]). After acidification the cylindrical CHEMS undergoes a change in the geometrical shape to a more inverted cone-shaped structure and thus promotes a hexagonal phase conversion together with the inverted cone-shaped lipid DOPE.

Amphoteric liposomes comprise charge reversible lipids containing pH-sensitive elements, such as the ionizable CHEMS (pK_α of ~5.8) and MoChol (pK_α of 6.5). CHEMS is thus an anionic lipid at physiological pH and the succinate moiety is protonated and uncharged at acidic pH. Vice versa, the morpholine moiety of MoChol is protonated at low pH and presents a cationic charge, at neutral pH the head group is deprotonated and non-charged. An equimolar mixture of both ionizable lipids (pK_α of ~6.3) is illustrated in Fig. 1.4.

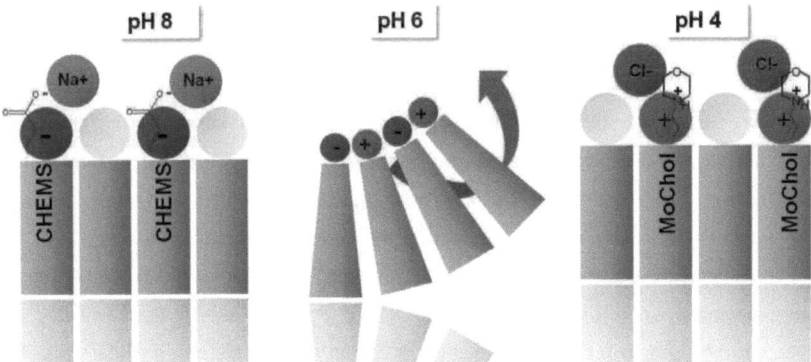

Fig. 1.4: Schematic illustration of an amphoteric membrane at different pH values. An equimolar mixture of ionizable lipids, CHEMS and MoChol, their counterion recruitment, molecular lipid shape and phase transition in dependency of the environmental pH is shown. Stable phases are important for oligonucleotide loading at low pH and storage and injection of the liposomes at physiological pH. Further details are noted in the text.

When charged, these lipids attract counterions which substantially increase head group volumes and thus promote the lamellar, non-fusogenic orientation.[125,126] Counterion binding causes a shape transition and stabilizes bilayers. Panzner and co-workers suggest that lipid bound counterions contribute a shape determining portion to the charged lipid and developed an extension of the lipid shape theory.[127,128] Transferred to a lipid bilayer, a stable phase occurs at ~pH 8 where MoChol is protonated and electrically neutral and the charged CHEMS recruits its counterion from the solvent. A second stable phase is at low pH where the charged MoChol recruits its counterion from the solvent and CHEMS is protonated and neutral. Fusion of the anionic CHEMS and the cationic lipid MoChol occurs in the absence of counterions at ~pH 6. The counterions are displaced and the charged lipids form an interlipid salt.[128] While displacing the counterions the volume of the head groups decrease which leads to a phase transition from stable lamellar to hexagonal phase. At this stage the membrane consisting of cone-shaped lipids and ion-free interlipid salt bridges is able to fuse with other membranes.[129]

The quantitative parameter describing the pH dependent phase behavior of lipid bilayers is the volume ratio between all polar and apolar elements and is called kappa (κ). The function implies the head group and tail volumes of anionic and cationic lipids, its molar fractions and was extended by the volume of the respective counterions.[127] The minimum kappa (κ_{min}) is calculated at the isoelectric point (IP) of the lipid mixture where the formation of an interlipid salt promotes a lipid phase transition. κ_{min} is thus the calculated volume ratio with highest tendency to fusogenicity.

It was shown by E.Siepi that large cationic counterions (e.g. Arginine, Tris) increase κ values and stabilize the anionic lipid bilayer more effectively. Conversely, membrane stability was diminished in the presence of small cationic counterions (e.g. Na) advantaged by small κ values. With respect to the stability and fusogenicity of cationic lipid bilayers similar results were found using different sizes of anionic counterions. Further, the impact of neutral lipids on the fusogenicity of a lipid mixture depends on their individual κ values. For example, POPC (κ =0.58) possesses a large head group and with increasing amount of POPC the κ_{min} value of the lipid mixture increases and fusion is reduced (determined by fusion assays [127]). Thus, POPC stabilizes membranes. In contrast, the small κ value of cholesterol (κ =0.09) decreases the κ_{min} value of the mixture and promotes fusion.

κ_{min} limits the transfection efficacy on HeLa cells. Liposomes with low values (<0.16) were substantially more effective than those with κ_{min} >0.3.[127] Mostly, the calculated κ_{min} values can be used as a predictive criterion for fusion and transfection of HeLa cells. The tight correlation between κ_{min} and transfection efficacy on primary mouse hepatocytes (PMHs) is demonstrated in this work. Therefore, the established ApoB100 model exemplified the transfection efficacy of liposomes *in vitro*, on PMHs, as well as *in vivo* in naïve mice.

1.7 ApoB100 – a valuable target to investigate oligonucleotide-mediated gene silencing *in vivo*

The apolipoprotein B100 (ApoB100) protein is present in plasma and is required for the assembly, secretion and the structural integrity of very-low-density lipoprotein (VLDL) and low-density lipoprotein (LDL) particles and acts as a ligand for the LDL receptor (LDLR) in various cells throughout the body. Elevated ApoB100 and LDL levels are associated with premature atherosclerosis in several inherited diseases, including familial hypercholesterolemia.[130] Abnormalities in ApoB100 metabolism are also observed in diabetes mellitus and obesity increasing the risk for coronary heart disease (CHD).[131,132]

This scaffolding protein ApoB100 is not amenable to conventional therapeutics such as small molecules, proteins, or monoclonal antibodies. Potential ASO and RNAi therapies have been developed against the so far "non-druggable" target ApoB100 which is predominantly expressed in the liver. Tab. 1.2 gives a comprehensive overview regarding the preclinical ApoB100 model targeted by oligonucleotides with or without delivery assistance. In the majority of the listed ApoB-trials, the wild type mouse strain C57Bl/6 was used, rarely BALB/c mice. In most cases ApoB-1 siRNA originally designed by Soutschek and co-workers [133] was selected for mouse trials. Within this work, C57Bl/6 mice were used for *in vitro* (isolated primary hepatocytes) and *in vivo* testing of formulated ApoB-1 siRNA or ApoB-ASO.

Company, Academia	Ref	Oligo	Delivery system	Dosing	Results (KD %)			additional analyses
					mRNA	protein	Chol:HDL:LDL	
ASOs								
ISIS	134	ISIS 147764	-	ip; 12 x 50mg/kg	L:88⁻	S: 90$^{\textit{\$}}$	66 : -- : 87	DR, AXT, histo, persist,
Santaris	57	SPC3716	-	3 x 5mg/kg	L:70⁻	--	85 : -- : --	Persist
siRNAs								
Alnylam	133	ApoB-1	Chol-conjugate	iv; 3 x 50mg/kg	L: 60*	P: 70⁺	40 : 25 : 40	BD, 5'-RACE
Alnylam, Protiva	135	ApoB-1	SNALPs	iv; 1 x 2.5mg/kg	L: 80*	S: 72⁺	-- : -- : --	DR, persist,
Alnylam, MIT	75	ApoB-2	Lipid-like conjugates	Iv: 1x 5mg/kg	L:70*	--	-- : -- : --	DR; persist hematology,
Mirus	136	ApoB-1	Dyn. Poly-conjugates	iv; 1 x 2.5mg/kg	L: 75⁻	S:~50$^{\textit{\$}}$	30 : -- : --	BD, histo, persist
Protiva	54	ApoB-1 (2'OMe)	SNALPs	iv; 3 x 5mg/kg	L: 80*	S:70⁺	50 : -- : --	Cytokines
RXi	137	ApoB	iNOPs	iv; 3 x 5mg/kg	L: 50⁻	P: 60$^{\textit{\$}}$	35 : -- : --	DR, cytokines
University of Tokyo	138	ApoB-1	α-Tocopherol	iv: 1x 2mg/kg	L:50⁻	--	20 : -- : --	persist, BD, DR, AXT, histo

Tab. 1.2: Setup and results of the mouse ApoB100 model: This table gives a comprehensive outline for various oligonucleotides and delivery systems in the preclinical mouse ApoB100 model. Abb.: KD: knockdown; analyses by * branched DNA; ⁺ ELISA; ⁻ qPCR; $^{\textit{\$}}$ WB; L: liver; P: plasma; S: serum; BD: biodistribution; AXT: liver enzymes ALT and AST; histo: histopathology; persist: persistence, DR: dose response; tox: toxicity analyses; ip: intraperitoneal; --: not applicable

1.8 Scope of the thesis

Chemical modifications have been used to facilitate the functional *in vivo* delivery of single-stranded ASO molecules, even in the absence of a delivery system. The number of organs or tissues that can be reached is, however, limited and insufficient delivery to more distal sites such as tumors and sites of inflammation is currently hindering the development of ASO inhibitors for such indications. The hurdles for systemic delivery of functional siRNAs into target cells are much higher and not outweighed by their higher potency. Therefore, assisted delivery of siRNA molecules is still a necessary condition for siRNA therapeutics.

The major objective of this thesis was to provide an amphoteric liposomal composition well-described by biophysical and pharmacological analyses for the effective delivery of oligonucleotides *in vitro* and *in vivo*.

The thesis aims first for the description and characterization of an amphoteric liposomal formulation (nov038) with a known ability to transfect macrophages and dendritic cells *in vivo*. The biodistribution and pharmacokinetic of nov038 encapsulating ASO molecules were investigated in a dose dependent manner to basically analyze the *in vivo* behavior of this formulation after systemic administration in mice. Based on these findings nov038 was prepared for pharmacodynamic studies demonstrating delivery of either therapeutic active ASO or siRNA molecules directed against parenchymal liver targets. Effective liposomal delivery of siRNAs requires the fusion with the endosomal membrane and the escape of the intact drug into the cytoplasm. In this context, the lipid composition of nov038 was shown to be non-fusogenic and thus inappropriate for the delivery of siRNAs.

The second part of the thesis aims on the creation and optimization of novel fusogenic liposomal compositions according to a rational design with a focus on the effective delivery of siRNA molecules. Prior to an *in vivo* use, these novel formulations were tested on primary mouse hepatocytes and a selected liposomal mixture (nov729) optimized for transfection efficiency and stability was further tested in the mouse ApoB100 model.

2. Materials and Methods
2.1 Materials

Chemicals and reagents used, unless noted otherwise, were purchased from Sigma-Alldrich (Schnelldorf, DE), Merck KGaA (Darmstadt, DE) and Roth (Karlsruhe, DE). Chemicals were of analytical grade and reagents were used according to the manufacturer's instructions. Buffer components were purchased from AppliChem (Darmstadt, DE) and Fluka (Seelze, DE) and were of molecular biology grade.

2.1.1 Lipids

All lipids used in this work are synthetic, HPLC purified and solvent free substances and were provided as dry powder. Lipids were purchased from the following manufacturers: cholesterol from Merck KGaA (Darmstadt, DE); CHEMS from Sigma Aldrich (Saint Louis, MO, USA); DOPE and POPC from Lipoid (Ludwigshafen, DE); DMGS and MoChol from Merck Eprova AG (Schaffhausen, CH); DODAP from Avanti Polar lipids (Alabaster, AL, USA).

Abb.	MW [g/mole]	Full name
	pK	Chemical structure
Tail vol. [Å^3]	Head vol. [Å^3]	
Chems	486.7	Cholesteryl-hemisuccinic acid
	5.39	
334.0	78.2	
Chol	387.0	Cholesterol
	14.90	
327.0	30.0	
DMGS	612.9	1,2-Dimyristoyl-sn-glycero-3-succinic acid
	5.33	
418.3	90.2	

Materials & Methods

Name	MW		Full name / Structure
DODAP	648.1		1,2-Dioleoyl-3-dimethylamino-propane (neutral form)
	7.52		
	511.8	45.7	
DOPE	744.0		1,2-Dioleoyl-sn-glycero-3-phosphatidyl-ethanolamin
	1.17 / 8.02		
	511.8	98.3	
MoChol	598.9		α-(3-O-cholesteryloxy)-δ-(N-ethylmorpholine)-succineamide
	6.51		
	334.0	168.2	
POPC	760.1		1-Palmitoyl-2-oleoyl-sn-glycero-3-phosphatidylcholine
	1.23 / 15.00		
	490.4	136.3	

Tab. 2.1: Lipid abbreviations & full name, structures, partial molecular volumes and pK values

2.1.2 Fluorescence Dyes

Name	MW [g/mole]	Chemical structure	Ex$_{max}$ [nm]	Em$_{max}$ [nm]	Supplier
Cy5.5	1128[#]		674	690	GE Healthcare UK Limited, Buckinghamshire, UK

Tab. 2.2: Fluorescence Dyes. [#] MW of Cy5.5 mono NHS ester

2.1.3 Oligonucleotides

Antisense Oligonucleotides (ASOs) targeting the surface receptor CD40 mRNA or LT1 and LT2 mRNA as well as a scrambled (scr, non-target control) ASO were a kind gift of ISIS Pharmaceuticals (Carlsbad, CA, USA). The sequence of scr ASO was taken for a 5' end modification using the fluorescence dye Cy5.5. This labeled scr-Cy5.5 ASO was ordered from NOXXON Pharma AG (Berlin, DE). The LNA-modified ApoB-ASO was taken from Swayze et al., 2007 and ordered from MWG Operon (Ebersberg, DE) with HPLC purity. All small interfering RNAs (siRNAs) were ordered from NOXXON Pharma AG (Berlin, DE). SiRNAs were provided in a desalted and HPLC purified form.

Name	MW [g/mole]	Sequence (5' → 3')	Species	Ref.
CD40	6391	CCCAgtcagtgttcCTGC	mouse	ISIS 117898*
scr	7152	CCTTCcctgaaggttCCTCC	mouse	ISIS 141923*
scr-Cy5.5	7500	Cy5.5-ccttccctgaaggttcctcc	mouse	-
LT1	7225	AGGTGctcaggactcCATTT	mouse	ISIS 101757*
LT2	7194	TCCATttattagtctAGGAA	mouse	ISIS217376*
LNA-ApoB	6619	**tc**tggtacatggaagtctgg	mouse	139

Tab. 2.3: Antisense oligonucleotides (ASOs): abbreviations, molecular weight, sequence and origin. Modifications: backbone of all ASO molecules are fully phosphorothiolated; capital letters: 2'MOE ribose; letters in bold: LNA ribose; Cy5.5: NIR fluorophore at 5'-end; * ISIS number representing a distinct ASO sequence in the respective species

Name	MW [g/mole]	Sequence: sense strand (5' → 3') antisense strand (5' → 3')	Species	Ref.
ApoB I	14974	guc auc aca cug aau acc aa*u	mouse /	133
		auu ggu auu cag ugu gau gaC* A*c	human	
ApoB I 5'P	15009	guc auc aca cug aau acc aa*u	mouse /	-
		P-auu ggu auu cag ugu gau gaC* A*c	human	
Scr (ApoB)	14973	gug auc aga cuc aau acg aa*u	mouse /	133
		auu cgu auu gag ucu gau caC* A*c	human	
scr-Cy5.5	15327	Cy5.5-aac ugg gua agc ggg cgc a-d(TT)	mouse /	140
		ugc gcc cgc uua ccc agu u-d(TT)	human	

Tab. 2.4: small interfering RNAs (siRNAs): abbreviations, molecular weight, sequence and origin. Modifications: asterisk: backbone phosphorothiolation; capital letters: 2'OMe ribose; **P**: 5'-phosphate; Cy5.5: NIR fluorophore at 5'-end; d(TT): two deoxy-thymidine overhangs

2.1.4 Antibodies

Name	Type	Reactivity	Specificity	Host	Supplier
ApoB (H300)	prim.	Human, mouse, rat	polyclonal	Rabbit	Santa Cruz Biotech., Inc Heidelberg, DE
GAPDH	prim.	Human, mouse, rat	monoclonal	Mouse	Abcam, Cambridge, UK
p38α	prim.	Human, mouse, rat	polyclonal	Rabbit	Abcam, Cambridge, UK
Alexa 680	sec.	Anti mouse	polyclonal	Goat	Invitrogen, Molecular Probes Carlsbad, CA, USA
IRDye 800	sec.	Anti rabbit	polyclonal	Goat	Rockland, Gilbertsville, PA, USA

Tab. 2.5: Primary (prim.) and secondary (sec.) antibodies.

2.1.5 Cells and Cell Culture

Description	Supplier
Mouse primary hepatocytes	see section 2.2.5.1
Antibiotics, penicillin & streptomycin (P/S)	Gibco, Invitrogen GmbH, Karlsruhe, DE
Bovine serum albumin (BSA)	Paesel & Lorei, Hanau, DE
Collagen R solution	Serva Electrophoresis GmbH, Heidelberg, DE
Collagenase NB4G	Serva Electrophoresis GmbH, Heidelberg, DE
DMEM (Dulbecco's Modified Eagle Medium)	Gibco, Invitrogen GmbH, Karlsruhe, DE
FCS (fetal calf serum)	PAA Laboratories GmbH, Pasching, AT
Mouse sera, aseptically filled	Sigma, St. Louis, USA
Optimem I (1x)	Gibco, Invitrogen GmbH, Karlsruhe, DE
TRITC-conjugated Phalloidin & DAPI	Millipore Corp., Bedford, MA, USA
RPMI 1640 Medium	Gibco, Invitrogen GmbH, Karlsruhe, DE
Serological Pipettes (5 ml, 10 ml, 25 ml)	TPP Ltd., Zurich, CH
Tissue culture test plates, 6-well-plates	TPP Ltd., Zurich, CH
Trypan blue, for cell culture	Sigma, St. Louis, MO, USA
Water (aqua destillata)	Gibco, Invitrogen GmbH, Karlsruhe, DE
In vivo-jetPEI™-Gal	PolyPlus-Transfection Inc., New York, NY, USA

Tab. 2.6: Primary hepatocytes, buffer and reagents for cell culture.

2.2 Methods

2.2.1 Preparation of Liposomes

2.2.1.1 Alcohol injection

Oligonucleotide-loaded Smarticles® were manufactured using the Advanced Loading Procedure (ALP) in which the interactions of cationic lipids (N) and the anionic phosphate backbone of the DNA or RNA oligonucleotides (P) are used to achieve a high payload into the liposomes. The loading process was performed at distinct N/P ratios (the Nitrogen/Phosphate molar ratio of cationic and anionic charge carrier) and under acidic condition (pH 4.0-5.0). The lipid mixture was dissolved to an appropriate concentration (initial lipid) in alcohol (ethanol, isopropanol or acidic isopropanol). Appropriate volumes of the oligonucleotide stock solution (according to the N/P ratio) were diluted in the respective acidic LOAD-buffer (Tab. 2.7). The organic and the aqueous solutions were mixed with two separately controllable pump systems at distinct flow rates to an alcohol content of either 10 % or 30 % (Tab. 2.7). Immediately, in the case of a 30 % alcohol injection the resulting liposomal suspension was diluted to a final concentration of 10 % alcohol by shifting to pH 7.5 with two times the volume of SHIFT-buffer (Tab. 2.7). Using the 10 % alcohol injection the pH of the liposomal suspension was shifted with 1/20 of the resulting total volume to pH 7.5 with the respective SHIFT-buffer (Tab. 2.7). Shifting of the pH value and / or salt concentration to physiological conditions diminished the interactions between oligonucleotides and lipids.

2.2.1.2 Concentration and separation

Formulations were concentrated using the tangential flow method and either MicroKros® hollow fiber membranes (Spectrum Labs, Inc., Rancho Dominguez, CA, USA) with a MW cut-off at 400 kDa and a surface area of 55 cm^2 or PelliconXL™ Biomax 100 PES cassettes (Millipore Corp., Bedford, MA, USA) with a MW cut-off at 300 kDa and a surface area of 50 cm^2, a Model 77201-60 Masterflex® easy-load® II, Console Drive pump, and Masterflex® 96440-16 tubing (Cole-Parmer Instrument, Vernon Hill, II, USA). During the concentration step the outside buffer, non-encapsulated oligonucleotides and organic solvent were exchanged by adding successively 7-times the volume of DIALYSIS-Buffer (Tab. 2.7). After the dialysis process and sterile filtration through 0.2 μm filter the liposomes adjusted to physiological pH and osmolarity were stored at a temperature of 2-8 °C.

Formulation	Alcohol type	alc. inj. [%]	N/P	Initial lipid [mM]	LOAD	SHIFT	DIA-LYSIS	TFM
nov038d144	Ethanol	10	4	20	#2	#2	#1	MK
nov038d145								
nov038d197			2.9				#2	
nov038d213			4.5	100			#1	P
nov038d222				120			#3	
nov038d231	Isoprop	30	2		#1	#1		
nov038d232							#1	MK
nov038d233			1.5	20				
nov038d234								
nov038d235			3				-	-
nov729d004			3.7					P
nov729d005								
nov729d017	Isoprop+	30	3.9	10	#3	#3	#1	
nov729d018								
nov729-Apo			4					MK
nov729-scr								
nov729-Cy55			3.9					

Tab. 2.7: **Critical production parameters for Smarticles formulations.** Abbreviations: alc. inj.: % alcohol injection; TFM: tangential flow method; P: PelliconXL PES cassettes; MK: MikroKros hollow fiber membranes; Isoprop+: acid Isopropanol incl. 25 mM CA

Formulation: nov038: POPC : DOPE : MoChol : Chems (15 : 45 : 20 : 20 mol %)

nov729: DODAP : DMGS : Chol (24 : 36 : 40 mol %)

LOAD-Buffer: #1 20 mM NaAc, 300 mM Sucrose, pH 4.0 (adjusted with HAc)
#2 20 mM HAc, 300 mM Sucrose, pH 4.5 (adjusted with Tris)
#3 10 mM CA, 280 mM Sucrose, pH 5.0 (adjusted with NaOH)

SHIFT-Buffer: #1 136 mM Na_2HPO_4, 100 mM NaCl, pH 9.0 (non-adjusted)
#2 1 M Tris, pH 8.0 (adjusted with HCl)
#3 100 mM Na_2HPO_4, 100 mM NaCl, pH 9.0 (non-adjusted)

DIALYSIS-Buffer: #1 PBS (Gibco), pH 7.4
#2 PBS (Na/K ratio), pH 7.4
#3 8.9 mM Na_2HPO_4, 3.0 mM KH_2PO_4, 280 mM Sucrose, pH 7.4

2.2.2 Characterization of Liposomes

2.2.2.1 Particle size determination

The particle size of liposomes was measured by dynamic light scattering using a 3000 HSA Zetasizer from Malvern Instruments Ltd. (Worcestershire, UK). Liposomes were diluted in PBS, pH 7.4 to a final lipid concentration of 0.1-0.2 mM. Average particle size is recorded as $Z_{average}$ value and size distribution (polydispersity index, PI) was calculated in the multimodal mode.

2.2.2.2 Determination of Zetapotential

The zeta potentials of liposomes were measured using a 3000 HSA Zetasizer from Malvern Instruments Ltd. (Worcestershire, UK). Liposomes were diluted in PBS, pH 7.4 or 10 mM HAc, 150 mM NaCl, pH 4.5 to a final lipid concentration of 0.04 mM. The zeta potentials were determined at both pH values.

2.2.2.3 Determination of lipid concentration

A) PHOSPHATE-Test: The inorganic phosphate concentration of final liposomal samples was determined according to van Veldhoven and Mannaerts, 1987 [141] and used as a measure of the total lipid concentration. This procedure, based on the complex formation of malachite green with phosphomolybdate under acidic conditions, was adapted to measure nanomolar amounts of phosphate, liberated from phospholipids after wet digestion.

B) CHOL-CHOD-PAP-Test: Lipid concentration of formulations without phospholipids was determined using the CHOL-CHOD-PAP-Test. The procedure bases on the enzymatic hydrolysis of cholesterol esters and the oxidation of cholesterol. The emerging hydrogen peroxide will be catalyzed to chinonimin in a peroxidase reaction. The colorimetric indicator chinonimin was measured photometrically at 546 nm. The CHOL-assay was conducted according to the manufacturer's instructions (Greiner Biochemika GmbH, Flacht, DE).

2.2.3 Determination of oligonucleotide concentrations

Oligonucleotide stock preparation: Lyophilized oligonucleotide samples were resuspended in 50 mM NaCl solution to a final concentration of approx. 10 mg/ml.

Oligonucleotides were dissolved for approximately 0.5 h at room temperature by repeated vortex mixing and sterile-filtered through a Minisart® filter with a pore size of 0.2 μm into a sterile Cellstar® test tube. Solutions were diluted 1:400 and 1:800 in 50 mM NaCl and the absorption intensity was determined at a wavelength of 260 nm on an UV/visible spectrophotometer according to: OD 1 = 40 μg/ml oligonucleotides
Oligonucleotide stock solutions were used immediately for experiments or stored long-term at a temperature of -70°C.

Determination of liposomal oligonucleotide: Oligonucleotide concentrations in the final liposomal suspension were photometrically determined from appropriate oligonucleotide standard curves. Lipids were extracted from samples in chloroform/methanol 1:1 (vol/vol) and the absorbance of the aqueous phase was determined at a wavelength of 260 nm on an UV/visible spectrophotometer.

2.2.4 Determination of non-encapsulated oligonucleotides

To verify the quality of the concentration and separation process the amount of non-encapsulated (outside) oligonucleotides within the final liposomal suspension was determined. Therefore, the formulations were diluted to a concentration of 30 ng/μl oligonucleotide with 20 mM Tris, 280 mM Sucrose, pH 7.4. An additional dilution (to 30 ng/μl) was prepared with 20 mM Tris, 280 mM Sucrose, pH 7.4, 1x loading buffer and 1 % Triton X-100 and incubated for 30 min at a temperature of 40 °C. During this treatment the liposomes were disintegrated and the encapsulated oligonucleotides were released from the liposomes. This Triton-treated sample served as a control for total oligonucleotide concentration.

A volume of 20 μl of those dilutions (600 ng oligonucleotides) was loaded onto a 15 % tris-borate EDTA (TBE) polyacrylamid gel. Free oligonucleotides were separated from encapsulated oligonucleotides for approx. 1 h at a voltage of 130 V. Further, a standard curve was prepared from the oligonucleotide stock solution and different amounts of oligonucleotide, e.g. 50 – 1000 ng in a volume of 20 μl, were loaded onto the gel.

The gel was stained with Stains-All working solution (10 ml stock stain (1 mg/ml Stains-All in formamide solution), 10 ml formamide, 50 ml isopropanol, 1 ml 3 M Tris, pH 8.8, 129 ml water) for 30 min in the dark with shaking and de-stained in distilled water under exposure to light for 30 min. The gel was scanned using the LI-COR Odyssey scanner (LI-COR Biosciences GmbH, Bad Homburg, DE) and stained bands were quantified with LI-COR application software. Band intensities were used to calculate the total and outside oligonucleotide concentration according to the prepared standard curve. The outside concentration was expressed as a percentage of total oligonucleotide concentration.

2.2.5 *In vitro* studies using primary mouse hepatocytes

2.2.5.1   Isolation of primary mouse hepatocytes

Hepatocytes from mouse livers were isolated according to a standard 2-step perfusion procedure.[142,143] Mice (strain C57Bl/6) were anaesthetized by 1.5 % isofluorane inhalation in O_2 at 2 l/min. The initial perfusion was conducted with 20 ml of Ca^{2+}-free Krebs–Ringer buffer made of 120 mM NaCl, 4.8 mM KCl, 1.2 mM $MgSO_4$, 1.2 mM KH_2PO_4, 24.4 mM $NaHCO_3$ and 250 mM ethylene glycol tetraacetic acid (EGTA) (pH 7.35) at 12 ml/min. Then perfusion buffer was changed to Krebs–Ringer buffer without EGTA but containing 15 mM HEPES (pH 7.5), 4 mM $CaCl_2$ and 0.75 mg/ml collagenase and perfusion was continued with 30 ml of buffer at 12 ml/min. The liver was excised, transferred to 20 ml of washing buffer (20 mM HEPES, 120 mM NaCl, 4.8 mM KCl, 1.2 mM $MgSO_4$, 1.2 mM KH_2PO_4, 0.4 % BSA, pH 7.4) and the dispersed cells were filtered through two layers of gauze to remove undigested material. The cells were then washed three times in washing buffer and sedimented each time at 50 g for 5 min at 4 °C.

2.2.5.2   *In vitro* transfection of primary mouse hepatocytes

For cultivation mouse hepatocytes were resuspended in 13 ml DME-Medium after the last washing step. Cell count and hepatocyte viability were determined with a hemocytometer after incubation with the non-viable-cells-indicator trypan blue. Hepatocytes were diluted to a final cell count of 2×10^5 living cells/ml with DME-Medium supplemented with 10 % FCS & 100 µg/ml P/S. Cultivation and *in vitro* transfection of hepatocytes were conducted on 6-well-tissue culture test plates. The 6-well-plates were pretreated and coated with a collagen/PBS-solution (0.5 mg/ml) for 30 min at 37 °C, washed with PBS (1x) and aqua dest (2x) and dried subsequently.

A total cell count of 4×10^5 hepatocytes in a volume of 2 ml were plated per well. The hepatocytes were cultivated in a humidified incubator at 37 °C and 5 % CO_2. Cells were washed with tempered PBS (2x) and supplied with fresh DME-Medium supplemented with 10 % FCS & 100 µg/ml P/S 24 h after plating and were transfected the next day.

Final liposomal suspensions with encapsulated oligonucleotides were diluted in Optimem I or the respective storage buffer to the appropriate concentration (11 times of the target concentration on the cells). For transfection a volume of 200 µl of the testing samples were added to 2 ml cell-surrounding medium by gently mixing (dilution factor of 11). Oligonucleotide concentrations tested on cells ranged from 1 to 1000 nM. Details of the tested lipid and oligonucleotide concentrations are summarized at the beginning of each *in vitro* study within the section "Results".

Saline or buffer (e.g. PBS) treated cells as well as untreated cells served as controls. Free, non-encapsulated oligonucleotides were transfected using the transfection enhancer *In vivo*-jetPEI™-Gal according to the manufacturer's instructions. For optimal complexation the transfectant and oligonucleotides were diluted in 0.1x PBS and oligonucleotide concentrations tested on cells ranged from 1 to 10 nM.

In the case of studies with supplemented mouse serum, a volume of 220 µl aseptically complete mouse serum was added to the cells (in 2 ml of DME-Medium) to a final concentration of 10 % (vol/vol) prior to the addition of the test samples.

After the treatment hepatocytes were cultivated for three days at 37 °C and 5 % CO_2 in a humidified incubator without a change of the cultivation medium. Afterwards the cells were prepared for mRNA analysis (see section 2.2.15.1 Quantigene).

2.2.6 Animal Trials

All animal trials except for "Pharmacodynamic of nov038-LT1 ASO" were conducted at Preclinics GmbH (Potsdam, DE) in accordance with animal care ethics approval and guidelines and were consistent with local, state and federal regulations as applicable (Landesamt für Verbraucherschutz, Landwirtschaft und Flurneuordnung, Referat Tierarzneimittel-Überwachung, Tierschutz, Frankfurt/O, DE). Naïve C57Bl/6 and NMRI mice (m/f) were purchased from Charles River Laboratories (Sulzfeld, DE). Mice were kept on a 12-h light/dark cycle with free access to food and water.

All test substances (e.g. saline, liposomal suspensions, buffered oligonucleotide solutions) were administered via tail vein injection. Details of injected volumes as well as lipid and oligonucleotide doses are summarized at the beginning of each animal trial within the section "Results". After dosing, if necessary, animals were anesthetized by isofluorane inhalation (a constant flow of 1-2 vol. % Forene® in pure oxygen) and blood was collected into EDTA-coated tubes by retrobulbar or heart bleeding. At the end of the study animals were sacrificed under isofluorane anesthesia; organ and blood samples were collected and prepared for subsequent analysis or stored at a temperature of -70 °C.

2.2.6.1 Pharmacokinetic (PK) and Biodistribution (BD) study

For the PK/BD study Smarticles formulation nov038d213 was loaded with a mixture of CD40- and scr-Cy5.5-ASO (4:1; w/w). The final liposomal suspension was serially diluted with PBS (Gibco) in order to administer different doses into mice. Studies were performed in 9 weeks old male NMRI mice (~35 g), grouped to a number of six (ID 1-3 and ID 4-6). Mice were treated once with either saline, non-encapsulated ASO or liposomal ASO using an injection volume of 250 µl. For pharmacokinetic purposes blood was collected by retrobulbar (0.5, 1, 2 and 4 h) and terminal bleeding (8 and 24 h) and analyzed according to the naïve pooling approach (Tab. 2.8). This modeling approach treats all data as coming from a single individual but without providing any estimate of interindividual variability. Organ samples of three mice were collected after 8 h and 24 h each and stored at a temperature of -70 °C immediately.

Pool	Mouse ID	0.5h	1h	2h	4h	8h	24h
PK	blood		retrobulbar blood			terminal bleeding	
A	1-3		X		X	X	
B	4-6	X	X				X
BD	organs			liver, spleen & kidney			
A	1-3					section	
B	4-6						section

Tab. 2.8: Blood and organ sampling for PK/BD study

2.2.6.2   Pharmacodynamic study in mouse liver ("nov038-LT1-ASO")

Mouse trial "nov038-LT1 ASO" was conducted at ISIS Pharmaceuticals (Carlsbad, CA, USA) and was in compliance with published Unites States Department of Agriculture regulations and approved by an Institutional Animal Care and Use Committee. Male 6-weeks old BALB/c mice (~25 g) from in-house breeding were grouped to a number of five and kept on a 12-h light/dark cycle with free access to food and water. Mice were injected iv with liposomal and non-liposomal samples twice a week for three weeks. The final liposomal suspension of nov038-LT1 was diluted 1:2 and 1:10 using PBS (Gibco) resulting in a dose of 0.25 mg (undiluted) 0.125 mg and 0.025 mg LT1 ASO per injection of 200 µl, which is the equivalent to a dosage of 10, 5 and 1 mg/kg, respectively. The mice were sacrificed 24 h following the last administration; liver samples were collected and prepared for LT1 mRNA (real time PCR) and protein (Western blot) analyses. Further, blood was collected by terminal bleeding and plasma was separated to determine AST and ALT levels.

2.2.6.3   Pharmacodynamic studies in mouse liver and plasma (ApoB100 trials)

Pharmacodynamic studies using either ApoB ASO or ApoB siRNA were performed in 8-10 weeks old C57Bl/6 mice (m/f, ~25 g), usually grouped to a number of five. Mice were intravenously injected either twice (day 1 and 3) or three times (day 1, 2 and 3) with the respective samples. Injections of saline served as a control. Mice were sacrificed on day 4 and prepared for organ and blood sampling. Liver tissue samples were used for the quantification of apoB100 mRNA using the Quantigene assay (2.2.15.1). Furthermore, whole blood was collected into EDTA-coated tubes by heart bleeding at the end of the study and plasma was separated by centrifugation. Plasma samples were prepared for the determination of ApoB100 protein using Western Blot analysis (2.2.16).

2.2.7 Total body and organ scan

In a satellite group within the PK/BD study one mouse was treated once with 4 mg/kg BW of liposomal Cy5.5-labeled ASO intravenously. The mouse was scanned 5, 15, 25, 45 and 60 minutes after injection of the liposomes using the LAS4000 luminescence scanner from Fujifilm (Fujifilm Corp., Tokyo, JP) at an emission wavelength of 670 nm. Therefore the mouse was anesthetized by isofluorane inhalation and prepared for a ventral body scan. One untreated mouse served as a blank control. Following the last scan after 60 min the mouse was sacrificed and organs (liver, spleen, kidney, lungs, heart and thymus) were collected and placed onto a petri dish for a separate organ scan using the same scanning parameters.

2.2.8 Determination of Cy5.5 fluorescence signals in blood samples (PK)

Blood samples from early time points were diluted 1:10 or 1:25 (0.5 h and 1 h), respectively, whereas samples from late time points were diluted 1:2 or 1:5 (2 h – 24 h), respectively. Dilutions were performed in PBS and 100 µl were transferred to a 96-well-plate. The plates were subjected to the LI-COR Odyssey Infrared Imaging System and fluorescence signals were densitometrically determined with the Odyssey Application Software Version 2.1. Background signal intensities of mouse blood treated with saline were subtracted from Cy5.5 treated samples. Average intensities were used to determine the Cy5.5 concentration (ng/ml) per sample according to an appropriate Cy5.5-ASO standard curve prepared in non-treated blood. Cy5.5 values (ng/ml) were multiplied with the respective assay dilution factor and the ASO-mixture factor of five (only 20 % of Cy5.5 labeled ASO per ASO dosage) resulting in the mean total ASO blood concentration [ng/ml; n = 3] per time point. The detection limit of Cy5.5 labeled ASO within the blood was at 4 ng/ml (20 ng/ml of total ASO), approximately. The total ASO blood concentration was plotted against the time and the following nonlinear regression fits (exponential decay) were calculated for mono- or bi-exponential distribution kinetics using the SigmaPlot 9.0 Software:

mono-exponential decay (one compartment): $y = a^*\exp^{(-d^*x)}$

bi-exponential decay (two compartments): $y = a^*\exp^{(-b^*x)} + c^*\exp^{(-d^*x)}$

Initial and terminal half-life ($t_{1/2}$) was calculated from b (ln2/b) and d (ln2/d), respectively, while the sum of the coefficients a and c represent the maximum ASO blood concentration C_{max}. Area under the ASO blood concentration-time curve (AUC) was performed using the log-linear integration method in that the concentration (C) monoexponentially declines between two measuring time points (t):

$$AUC_{t1-t2} = ((C_1-C_2)/(\ln(C_1/C_2)))*(t_2-t_1)$$

The calculated C_{max} served as the blood concentration at time point t_0 and the AUC was extrapolated to infinity by dividing the last plasma concentration by the slope of the terminal phase (C_{last}/d). Total body clearance (CL_{tot}) was calculated by dividing the ASO dose by the measured AUC: $\qquad CL_{tot} = ASO\ dose\ /\ AUC_{t0-\infty}$

The remaining ASO blood concentration per time point, expressed as % of injected dose, was calculated using a blood volume of 2.7 ml per mouse and referred to the respective injected dosage.

2.2.9 Determination of Cy5.5 fluorescence signals in tissue samples (BD)

For determination of total amount of Cy5.5 labeled ASO per organ a piece (approx. 50 mg) of either liver, spleen or kidney were homogenized in 250 µl of homogenization buffer (taken from Quantigene 1.0 Reagent System) using a vibratory mill. Organ homogenates were diluted 1:5 and 1:10 in PBS and 100 µl each were transferred to a 96-well-plate. The plates were subjected to the LI-COR Odyssey Infrared Imaging System and fluorescence signals were densitometrically determined with the Odyssey Application Software Version 2.1. Background signal intensities of mouse tissues treated with saline were subtracted from Cy5.5 treated samples. Average intensities were used to determine the Cy5.5 concentration (ng/ml) per sample according to an appropriate standard curve made from Cy5.5-ASO stock solution diluted in PBS. Cy5.5 values (ng/ml) were multiplied with the respective dilution factor and the total amount per organ was calculated. For calculation purposes a mean liver, spleen and kidney weight of 2085 g, 148 g and 282 g was used, respectively. The total amount of labeled material per organ and per time point was determined from three animals in duplicate and expressed as "% of injected dose".

2.2.10 Cryosections

Frozen organ samples were partially embedded in Tissue-Tek® (O.C.T.) and 10 µm sections were cut using a Cryostat Cryo-Star HM560 (Microm-International, Walldorf, DE) at -20 °C. Sections were placed onto Super Frost® Plus Gold slides and stored at 4 °C.

2.2.11 Fluorescence microscopy

Organ sections were fixed in a methanol: acetone solution (1:1, vol/vol, 5 min) and washed in PBS. The sections were then counterstained with 4',6-Diamidino-2-phenylindol (DAPI) solution (1 µg/ml in PBS, 10 min) and washed again in PBS. Cryosections were mounted with Citifluor AF1 and capped with a cover slip. Cryosections were examined using an Axiovert100 microscope (Carl Zeiss GmbH, Jena, DE) with an HBO100 mercurial lamp and appropriate fluorescence filters. Images were taken using a monochrome CCD-Camera without IR-Filter. Exposure times for all sections at the highest magnification (200x) were: 2 s for Cy5.5; ~0.1 s for DAPI. The monochrome images were edited with the Zeiss AxioVision 4.0 software.

2.2.12 Confocal Laser Scanning Microscopy (CLSM)

For CLSM mouse primary hepatocytes were supplied with fresh DME-Medium / 10 % FCS and cultivated in 24-well-plates on etched (40 % HCl, 60 % EtOH) cover slips with ~5x10^4 cells/well in a humidified incubator at 37 °C and 5 % CO_2. The cells were transfected with Cy5.5 labeled ASO or Cy5.5 labeled siRNA encapsulated into liposomes. After 4 h of incubation cells were washed twice with PBS and supplied with fresh DME-Medium + 10 % FCS. 24 h following the transfection step cells were washed once with PBS. Subsequently the cells were fixed in 4 % paraformaldehyde (PFA) for 20 min. Then the cells were counterstained with Phalloidin-TRITC (1:1000) and DAPI (1:50000) diluted in PBS (both from Millipore Corp., Bedford, MA, USA) for at least 10 min. Afterwards the cells were washed twice with distilled water and twice with ethanol (96 %). The cells were dried, mounted with Citifluor AF1 and capped with a cover slip.

Untreated liver sections were fixed in 4 % PFA for 30 min. DAPI counterstaining, washing, dehydration, drying and mounting of the samples was proceeded as described above.

Confocal images were taken using a LSM SP5 from Leica Microsystems (Wetzlar, DE) with standard parameters for sequential image acquisition. The excitation was set according to the used fluorescence dyes (DAPI 405 nm; TRITC 561 nm and Cy5.5 633 nm).

2.2.13 Determination of plasma values (liver enzymes and plasma cholesterol)

Whole blood was collected into EDTA-coated tubes, and plasma was separated by centrifugation. For analysis of total cholesterol, HDL-Chol, LDL-Chol, AST and ALT a plasma volume of 300 μl was sent to the Institut für Veterinärmedizinische Diagnostik GmbH (Berlin, DE). Plasma parameters were determined using a clinical analyzer of Roche / Hitachi. All analysis reagents (Cobas® series) were purchased from Roche Diagnostics GmbH (Mannheim, DE). Activities of plasma enzymes AST and ALT were photometrically determined after enzymatic conversion reactions. Cholesterol levels were assayed by using an enzymatic colorimetric method according to the CHOL-CHOD-PAP-Test (2.2.2.3 B).

2.2.14 Cytokine ELISA

For the quantitative measurement of four cytokines (IL-1ß, IL-6, IFNγ and TNFα) in plasma samples the Searchlight® IR Mouse Cytokine Array from Pierce Biotechnology, Inc. (Rockford, Il, USA) was used. This multiplex sandwich ELISA was provided with specific antibodies spotted on each well and capturing specific cytokines in the standards and samples added to the plate. The biotinylated detecting antibodies were added, a Streptavidin-DyLight800-conjugate mediated the specific binding to Biotin and infrared fluorescence signals were measured with the LI-COR Odyssey reader. The amount of signal produced in each spot is proportional to the amount of cytokine in the original standard or sample. Plasma samples were diluted 1:5 using the provided Serum/Plasma Sample Diluent and the assay was performed according to the manufacturer's instructions. Total amounts of cytokines [pg/ml] were calculated from an appropriate standard curve provided by the manufacturer.

2.2.15 Quantification of mRNA

2.2.15.1 QuantiGene (QG)

QuantiGene Reagent System (Affymetrix, Fremont, CA, USA) is a sandwich nucleic acid hybridization assay performed on 96-well plates that provides an approach for RNA detection and quantification by amplifying the reporter signal using branched DNA (bDNA) technology.

The ability to quantify specific RNA molecules within a sample lies in the design of a QuantiGene Probe Set. Each oligonucleotide probe set contains three types of synthetic probes, Capture Extenders (CEs), Label Extenders (LEs), and Blockers (BLs) that hybridize to contiguous sequences of the target RNA. The CEs bind to the capture oligonucleotides conjugated to the well surface and capture the associated target RNA via cooperative hybridization (Fig. 2.1: Step 1). Signal amplification is mediated by DNA amplification molecules that hybridize to the tails of the LEs (Fig. 2.1: Step 2). Each amplification unit contains hybridization sites for multiple alkaline phosphatase (AP) conjugated Label Probes, which can then be detected by the AP mediated degradation of a chemiluminescent substrate (Fig. 2.1: Step 3). Luminescence is reported as relative light units (RLUs) on a microplate luminometer. The amount of luminescent signal is linearly proportional to the number of RNA molecules present in the sample.

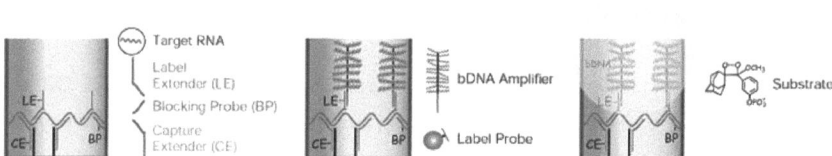

Fig. 2.1: Principles of QuantiGene Assay. Figures are taken and modified from the QuantiGene Reagent System User Manual.

ApoB100 mRNA levels were quantified using the QuantiGene 1.0 Reagent System. For normalization of apoB100 expression across all samples the housekeeping gene peptidylprolyl isomerase B (PPIB, cyclophilin B) was used. For detection of mouse mRNA, the ApoB100 probe set was specific to mouse ApoB (positions 5183-5811, XM_137955) and the PPIB probe set was specific for mouse PPIB (positions 24-522, NM_011149).

A) ApoB100 mRNA analysis in tissue samples: After collection of the liver tissue small uniform pieces (approx. 20 mg) were excised from one liver lobe and transferred to 1 ml of RNAlater (Ambion, Inc., Austin, TX, USA). The liver samples were incubated at RT for 4-6 h and afterwards for 24 h at 4°C. Thereafter liver samples were stored at a temperature of -70°C. Frozen liver samples from mice were thawed and approximately 10 mg were transferred to 250 µl homogenization buffer containing proteinase K at a concentration of 0.2 mg/ml. For homogenization purposes one steel bead was added to the liver sample and the tissue was homogenized using a vibratory mill at a beat frequency of 25/s for 2 min. The homogenization step was repeated until no tissue pieces were visible. After homogenization the steel beads were removed from the reaction tube and liver lysates were incubated at 65°C for 0.5 h. Cell debris were separated by centrifugation at 13.000 rpm for 10 min. The supernatant was diluted (factor 30) in diluted lysis mixture (DLM) and processed in a 96-well plate according to the manufacturer's instructions. The luminescence was determined on a FLUOstar OPTIMA reader (BMG Labtech GmbH, Offenburg, DE) at a gain level of 3000 and a measurement interval time of 0.5 sec.

B) ApoB100 mRNA analysis in cell lysates: The medium was discarded and hepatocytes were washed with PBS twice. The cells were lysed in diluted lysis mixture (DLM, 1 ml per well) containing proteinase K at a concentration of 50 µg/ml. The tissue plates were incubated and gently shook for 30 min at a temperature of 37 °C. After cell lysis the samples were transferred to reaction tubes. The cell lysates were diluted (factor 3) in DLM and processed in a 96-well plate according to the manufacturer's instructions. The luminescence was determined on a FLUOstar OPTIMA luminescence reader at a gain level of 3000 and a measurement interval time of 0.5 sec.

2.2.15.2 Real-time PCR

The quantification of LT1 mRNA levels in whole liver lysates was performed by ISIS Pharmaceuticals (Carlsbad, CA, USA). Liver tissues were homogenized in guanidinium isothiocyanate followed by cesium chloride gradients and total RNA was extracted with Qiagen RNeasy isolation kits. An RNA amount of 50 ng of each sample was subjected to RT-PCR analysis; all reagents were from Invitrogen (Carlsbad, CA, USA). Mouse-specific primer pairs and probes (Tab. 2.9) were used for the quantification of LT1 mRNA and values were normalized to glyceraldehyde-3-phosphate dehydrogenase (GAPDH). In both cases, the probes were labeled with 5'-FAM (6-carboxyfluorescein reporter) and 3'-TAMRA (5(6)-carboxytetramethyl-rhodamine quencher).

Primer	LT1	GAPDH
Forward	AAGGGAACGAGAAAACTGCTGTT	GGCAAATTCAACGGCACAGT
Reverse	TATTTTAACCAGTGGTATTATCTGACATCCT	CGCTCCTGGAAGATGGTGAT
Probe	*FAM*-TTGTATTTGTGAACTTGG-CTGTAATCTGGTATGCC-*TAMRA*	*FAM*-AAGGCCGAGAATG-GGAAGCTTGTCATC-*TAMRA*

Tab. 2.9: Primer sequences for real-time PCR of LT1 and GAPDH mRNA. All primer sequences are given in 5' → 3' orientation.

2.2.16 Western Blot Analysis

2.2.16.1 Sample Preparation for protein analysis

A) LT1 protein: Frozen liver samples from mice treated with free antisense or antisense encapsulated into Smarticles formulation nov038 were sent by Isis Pharmaceuticals

(Carlsbad, CA, USA). Aliquots of 100 mg were cut frozen and homogenized on ice in 1 ml of lysis buffer (20 mM Tris-HCl, 150 mM NaCl, 5 mM $MgCl_2$, 1 mM EDTA, 1 mM phenylmethylsulfonylfluorid (PMSF)) using a Dounce homogenizer. After homogenization the samples were transferred to 2 ml reaction tubes. Triton X-100 was added to the homogenates at a final concentration of 10% and the lysates were incubated on ice for 15 to 30 min. Cell debris were separated from homogenates by centrifugation at 13000 rpm for 10 min at a temperature of 4°C. Total protein concentrations were determined in supernatants using the bicinchoninic acid assay (BCA) according to the manufacturer's instructions and adjusted to a concentration of 1 µg/µl protein in sample buffer. Aliquots of liver homogenates from all mice in each treatment group were pooled. The remaining volumes of homogenates were stored at a temperature of -70°C.

B) ApoB100 protein: For Western Blot analysis volumes of 50 µl plasma were mixed with 50 µl 4x NuPage® LDS sample buffer containing 1x protease inhibitor cocktail. Samples were heated to a temperature of 65 °C, stored for 4 h at a temperature of 4 °C, and subsequently frozen at -70 °C. Denatured plasma samples were further diluted 1:2.5 in 4x LDS sample buffer containing 1x sample reducing agent and heated again to a temperature of 65 °C for 10 min to resolve small precipitates and to maintain the denatured protein form.

2.2.16.2 Western Blot analysis

Protein	LT1	ApoB100
I. Samples		
	Liver lysates, see 2.2.8.1 A)	Plasma, see 2.2.8.1 B)
II. Gel loading		
	~10 µg total protein	2-3 µl of total plasma
III. Electrophoresis		
Gel type	SDS-PAA 10 %, Bis-Tris, Novex	SDS-PAA 3-8 %, Tris-Acetate, Novex
Running buffer	1x MOPS, SDS, NuPage	1x Tris-Acetate SDS, NuPage
Voltage [V] // Current [mA]	200 // 150	100 // 30-40
Running time [h]	1	1.5
IV. Blotting		
Type of Blotting	Semi Dry onto PVDF membrane	Semi Dry onto PVDF membrane
Transfer buffer	10 mM CAPS, pH 11 (NaOH), +10 % MeOH	2x NuPage, +10 % MeOH, + AntiOxidant
Voltage [V] // Current [mA]	5 // 400	2-4 // 300
Running time [h]	1.5	3

V. Blocking		
Blocking reagent	2 % Block reagent (Amersham) in TN + Tween (0.1 %) buffer	Odyssey Blocking Reagent
Running time [h]	16 (o/n)	2
Temperature [°C]	4	RT
VI. Primary Antibody		
Name	Anti mouse p38α-Ab; rabbit IgG // Anti mouse GAPDH-Ab; mouse IgG	Anti mouse ApoB100-Ab; rabbit IgG
Dilution	1:1000 // 1:3000	1:200
Buffer	2 % Block reagent in TN buffer	Odyssey Blocking Reagent
Running time [h]	2	2
Temperature [°C]	RT	RT
Washing buffer // Time	TN + Tween (0.1 %) // 3x 10 min	1x PBST // 3x 10 min
VII. Secondary Antibody		
Name	Alexa680-anti-mouse-IgG; goat-IgG // IRDye800-anti-Rabbit-IgG; goat-IgG	IRDye800-anti-Rabbit-IgG; goat-IgG
Dilution	Each 1:5000	1:15000
Buffer	2 % Block reagent in TN buffer	Odyssey Blocking Reagent
Running time [h]	2	1
Temperature [°C]	RT	RT
Washing buffer // Time	TN + Tween (0.1 %) // 3x 10 min	1x PBST // 3x 10 min
VIII. Detection		
	LI-COR Odyssey NIR scanner, 700 and 800 nm channels	LI-COR Odyssey NIR scanner, 800 nm channel

Tab. 2.10: Detailed descriptions for the conduction of LT1 and ApoB100 Western blot analysis. TN: Tris/NaCl buffer; taken from: Abcam (www.abcam.com/technical); PBST: Phosphate buffered saline + 0.2 % Tween 20; taken from: Abnova (www.abnova.com.tv)

Staining intensities of protein bands were densitometrically determined using the LI-COR near infrared scanner and the Odyssey 2.1 software. The amount of LT1 protein was determined from the ratio of the integrated intensity of the stained band to that of the internal standard GAPDH. Sample preparation for LT1 protein analysis and LT1 Western Blot analysis were performed by Drs. Ludger Ickenstein and Evgenios Siepi.

2.2.17 Statistical analyses

Data are expressed as means ± SD. Statistical significance of differences was determined using a 2-tailed Student's *t*-test assuming equal variance. P values <0.05 were considered statistically significant.

3. Results

This work aimed at the development of novel liposomal formulations encapsulating different types of oligonucleotides (ASO and siRNA) and to basically describe their functional behavior *in vivo*. For that, biodistribution and pharmacokinetic as well as pharmacodynamic studies were performed in mice to understand the blood circulation, distribution and accumulation of specific Smarticles formulations in distinct tissues. Further, the sequence-specific knockdown of a target mRNA and protein indicates the potency of these liposomes delivering oligonucleotides into tissue cells.

3.1 Pharmacokinetics and Biodistribution of nov038

In a series of studies, the circulation and deposition of an antisense molecule encapsulated into Smarticles formulation nov038 was followed. It is known that the distribution of liposomes is dose dependent and, typically, two kinetic compartments are described in the literature, wherein at least one can be saturated.[88] It was thus a main objective of this study to analyze the distribution and pharmacokinetic of nov038 loaded with Cy5.5-labeled ASO in a dose-dependent manner.

Batch production and study parameters for nov038 are presented in Tab. 3.1. Nov038 was loaded with a mixture of CD40- and scr-Cy5.5-ASO (4:1; w/w) resulting in an average particle size of 174 nm. The liposomal suspension was prepared to a final concentration of 1054 µg/ml ASO and 124 mM lipid leading to a drug-to-lipid ratio of 8.5 µg ASO/µmol lipid.

	Saline	Free Cy5.5-ASO	Nov038-Cy5.5-ASO			
Lot#	-	-	d213			
Ave. particle size [nm] / PI	-	-	174 / 0.18			
Total ASO conc. [µg/ml]	-	280	1054	527	264	132
Non-encaps. ASO [%]	-	-	13			
Lipid conc. [mM]	-	-	124	62	31	16
Drug-to-lipid ratio [µg/µmol]			8.5			
Injection volume [µl]	250	250	250			
ASO dose [mg/kg]	-	2	6.5	3.3	1.6	0.8
Lipid dose [µmol/kg]	-	-	886	443	222	111

Tab. 3.1: Nov038d213 sample and study parameters. The dose-dependent PK and BD of nov038 were followed by using an encapsulated Cy5.5-labeled ASO. Nov038d213 was serially diluted to achieve four different dosages *in vivo* which were calculated from encapsulated ASO.

3.1.1 Whole body imaging indicates a fast distribution into liver and spleen

First, the overall biodistribution of nov038-Cy5.5-ASO was monitored in mice by whole body imaging. The final liposomal suspension of nov038d213 was diluted by a factor of 2 leading to a concentration of 527 µg/ml ASO and 62 mM lipid. A single dose of 4 mg/kg encapsulated ASO and 532 µmol lipid/kg BW was intravenously administered. The biodistribution of the labeled material was followed over 1 h using a Fujifilm LAS4000 luminescence scanner (Fig. 3.1). Previous internal BD-studies showed that 10-50 µg (0.4-2 mg/kg) of a Cy5.5-labeled ASO are sufficient to follow the NIR-dye within the body.

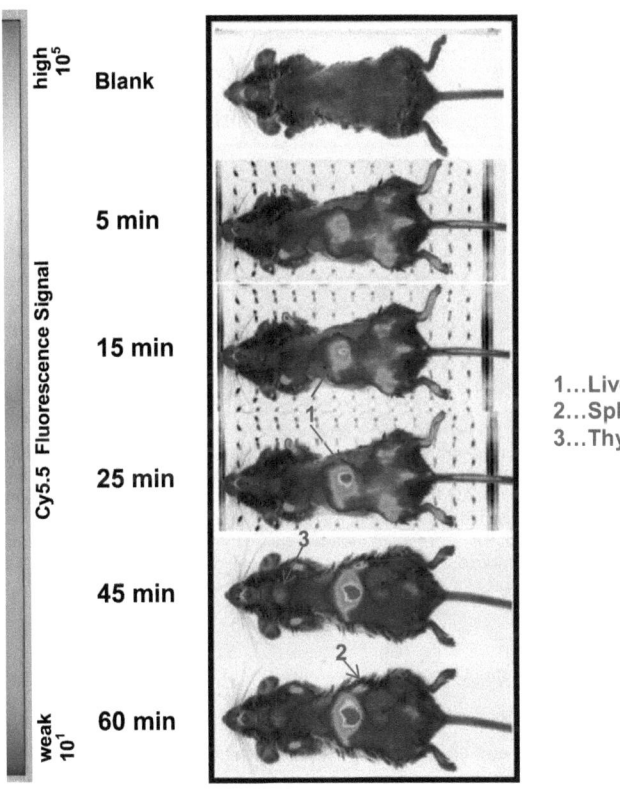

1...Liver
2...Spleen
3...Thymus

Fig. 3.1: Whole body imaging after a single treatment with nov038 encapsulating Cy5.5-labeled ASO using a Fuji Film LAS4000 scanner at a wavelength of 670 nm. The figure shows ventral scans of the treated mouse after 5, 15, 25, 45 and 60 minutes and one untreated mouse which served as a blank control. Blue signals indicate a weak fluorescence whereas yellow-to-red colored tissues demonstrate a high uptake of the Cy5.5-labeled ASO. A fast invasion of nov038-Cy5.5-ASO into liver and spleen is observable whereas no material is detectable in lungs and heart.

A non-treated mouse served as a blank control and no specific Cy5.5 fluorescence signals were visible after whole body imaging (Fig. 3.1). As expected the mouse treated with nov038-Cy5.5-ASO showed a high uptake of the fluorescent material into liver and spleen over the time. The spleen is dorsally displaced and hence not very obvious after a ventral body scan. Also, the uptake of the material into the thymus was visible and tissues, e.g. paws and snout, which are well supplied with blood, accumulated the liposomal Cy5.5-labeled ASO to some extent. Due to a close proximity of the kidneys to the liver overwhelming any other tissue signals no distinct kidney signals were visible in the whole body scan. No signal was detectable in the heart and only weak fluorescence signals could be found in the lungs. Potential Cy5.5 signals were not covered by the thoracic bones because a separate scan from dissected organs revealed an almost unstained lung and heart tissue. Further, a weak uptake of the Cy5.5 label into the kidneys was confirmed by this separate organ scan (Fig. 3.2).

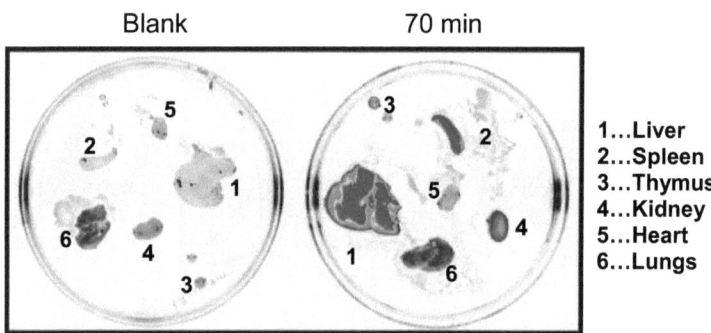

Fig. 3.2: Organ scan of an untreated mouse (Blank, left petri dish) and a nov038-Cy5.5-ASO treated mouse 70 minutes after injection (right petri dish). Liver, spleen, thymus, kidney, heart and lungs were excised and placed onto a petri dish. No Cy5.5 fluorescence signals were visible in the organs of the control mouse whereas high uptake of the fluorescent material was confirmed in liver and spleen of the liposomal treated mouse. Further, weak signals could be detected in the kidney and thymus whereas no or low signals were detectable in the heart and lungs.

Macroscopically, nov038 with encapsulated Cy5.5-labeled ASO predominantly distributed into liver and spleen which was confirmed by organ scans. The accumulation of the labeled material in liver and spleen took place to equal local concentrations indicated by pink-colored areas. The liposomes distributed further to the thymus whereas no or less signals were visible in the heart and lungs. A turquoise colored kidney indicates a weak uptake of the, most likely non-encapsulated, labeled material.

In addition, the NIR properties of Cy5.5 ($\lambda_{ex}/\lambda_{em}$ = 674/690 nm) make it suitable for *in vivo* applications, which has already been shown in previous reports.[144,145]

3.1.2 Pharmacokinetic of free and encapsulated Cy5.5-labeled ASO

Serial dilutions of the nov038d213 "stock" solution were prepared resulting in four liposomal test samples with decreasing ASO and lipid concentration. As mentioned in Tab. 3.1 mice received a single *in vivo* dose of either 6.5 mg, 3.3 mg, 1.6 mg or 0.8 ASO/kg BW with 886 µmol, 443 µmol, 222 µmol or 111 µmol lipid/kg BW, respectively. A saline treated group and free, non-encapsulated ASO at a dose of 2 mg/kg served as controls.

To investigate the pharmacokinetic behavior of free, non-encapsulated ASO and liposomal ASO blood was collected after distinct time-points and the NIR-signal was quantified. The pharmacokinetic profile of free ASO and nov038-Cy5.5-ASO injected at different doses in mice is shown in Fig. 3.3.

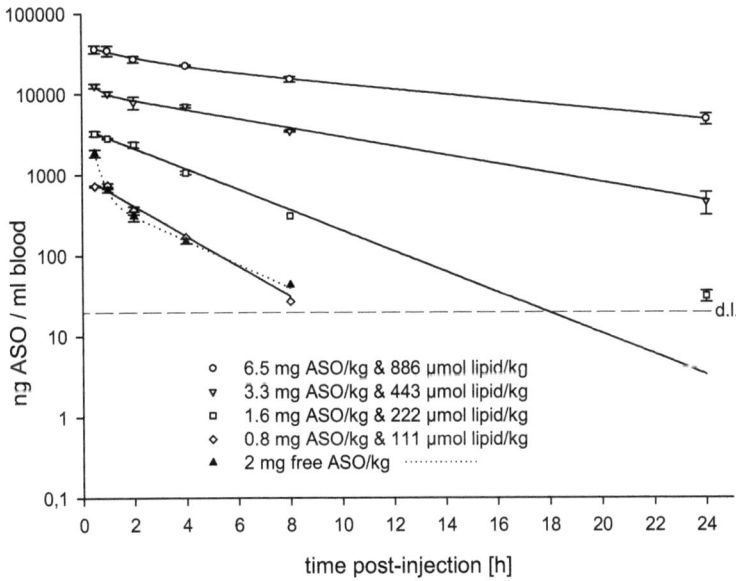

Fig. 3.3: Pharmacokinetic profile of free ASO and nov038-Cy5.5-ASO in mice. The blood concentration of Cy5.5-labeled ASO was determined using a fluorescence reader and the decrease of the total ASO blood concentration (ng/ml) over the observation period [24 h] is shown. Data points represent the mean [n = 3] ± SEM and were fitted using an exponential decay regression. The coefficient of determination, r^2, indicated a good approximation of the regression fit to the experimental data in all groups ($r^2 \geq 0.985$). Circulation times of nov038 depended on the injected dose. By increasing the lipid (and ASO) dose the blood half-life of the liposomes was prolonged. Free ASO is cleared from the blood stream very rapidly. d.l. = detection limit

High lipid doses of nov038 with encapsulated Cy5.5-labeled ASO showed a biexponential pharmacokinetic profile (Fig. 3.3) which is characterized by a first very fast decrease of the ASO blood concentration (distribution phase) and a second much slower elimination phase. Within the first 30 minutes most of the material (>60 %) has been extravasated from the blood stream (Tab. 3.2), presumably into the primary organs liver and spleen, as shown in Fig. 3.1. The following PK is characterized by dose-dependent elimination of the drug carrier from the blood stream. A blood concentration of 4860 ng ASO / ml (5 % of injected dose) was still detectable after 24 h in mice treated with the highest dose of nov038-Cy5.5-ASO. This referred to an elimination (terminal) $t_{1/2}$ of ~10 h (Tab. 3.3).

(% of injected dose)	Free ASO 2 mg/kg	Nov038-Cy5.5-ASO (ASO / lipid dose)			
		6.5 mg/kg 886 µmol/kg	3.3 mg/kg 443 µmol/kg	1.6 mg/kg 222 µmol/kg	0.8 mg/kg 111 µmol/kg
0.5 h	7.2	36.7	25.9	13.3	5.9
1 h	2.5	35.3	20.8	11.6	6.1
2 h	1.2	27.6	16.1	9.7	3.0
4 h	0.6	22.9	14.5	4.3	1.4
8 h	0.2	15.5	7.1	1.3	0.2
24 h	-	5.0	0.9	0.1	-

Tab. 3.2: Relative blood levels (expressed as % of injected dose) of free, non-encapsulated ASO and different doses of nov038-Cy5.5-ASO. Amounts of Cy5.5-labeled ASO within the blood (ng ASO / ml blood, Fig. 3.3) were calculated as percentage of injected dose. The amount of labeled ASO clearly decreased over the observation period in a dose-dependent manner. In the groups of free ASO (2 mg/kg) and nov038-Cy5.5-ASO injected at dose of 0.8 mg ASO/kg no Cy5.5-labeled ASO was detectable after 24h.

Relative blood levels (% of injected dose) at distinct time points and terminal $t_{1/2}$ decreased with lowering the dose. At a dose of 0.8 mg/kg of nov038-Cy5.5-ASO only 6 % of the injected dose was recovered in the blood stream after 0.5 h. Almost 100 % of encapsulated ASO distributed to the primary organs within the first minutes (see section 3.1.3) and no material was detected within the blood stream after 24 h (Tab. 3.2). Because most of the injected material was cleared from the blood stream during the distribution phase a single exponential regression fit was used leading to only one calculated blood half-life of 1.6 h (Tab. 3.3). Generally, all groups show a massive initial decay of the ASO blood concentration (Tab. 3.2) which indicates a very voluminous first compartment and/or a much faster distribution phase within the first 30 minutes. The C_{max} values, calculated by exponential regression fits, indicate the maximum concentration after injection based on the experimental data. While the bisection of high lipid doses of nov038 (886 → 443 µmol lipid/kg) led to a decay of C_{max} by a factor of two (~40500 → ~21000 ng ASO/ml), further 2-fold lipid reduction (443 → 222 → 111 µmol lipid/kg) resulted in a C_{max} decay of a factor 4-5 (Tab. 3.3). Further, as the dose increased, the area under the ASO blood concentration-curve (AUC) increased in a dose-related, but not dose-linear fashion (Tab. 3.3). The same applies for the total body clearance (CL_{tot}) which decreases by a constant factor of approx. three (15 → 4.9 → 1.6 → 0.7) with doubling the doses. The rate constants in both, the distribution and elimination phase strongly depends on the lipid dose. Interestingly, by doubling the lipid (and ASO) dose the terminal $t_{1/2}$ doubled also (1.6 - 2.4 - 5.4 - 9.8 h; Tab. 3.3). In summary, the collected data and calculated parameters indicate a non-linear PK for nov038 which can be described by a two-compartment model.

	dose [mg ASO/kg] / [µmol lipid/kg]	C_{max} [ng ASO/ml]	initial $t_{1/2}$ [h]	terminal $t_{1/2}$ [h]	$AUC_{t0-\infty}$ [µg ASO/ml * h]	CL_{tot} [ml/h]
Free ASO	2 / -	9090	0.2	2.0	4.2	16.59
Nov038-Cy5.5-ASO	6.5 / 886	40462	1.4	9.8	403.8	0.65
	3.3 / 443	21025	0.3	5.4	85.2	1.55
	1.6 / 222	3848	-	2.4	13.6	4.85
	0.8 / 111	974	-	1.6	2.2	15.01

Tab. 3.3: Pharmacokinetic parameters of free ASO and different doses of nov038-Cy5.5-ASO. The maximal ASO concentrations in the blood, C_{max}, as well as the initial and terminal half-life ($t_{1/2}$) were calculated from exponential regression fits of experimental collected data. Data points of free ASO and the two highest doses of nov038-Cy5.5-ASO were fitted according to a bi-exponential decay regression. The lower doses of nov038-Cy5.5-ASO were fitted using a mono-exponential decay regression. All PK parameters including the Area under the ASO blood concentration-time curve (AUC) and the total body clearance (CL_{tot}) reveal a missing dose proportionality indicating a non-linear PK behavior for nov038.

Fig. 3.3 and Tab. 3.2 clearly show a high elimination and rapid blood clearance of free, non-encapsulated ASO characterized by an initial $t_{1/2}$ of 0.2 h and a terminal $t_{1/2}$ of 2 h in the blood circulation (Tab. 3.3). Supportingly, a low AUC (4.2 µg ASO/ml * h) and a comparatively high clearance rate (16.6 ml/h) were calculated for a dose of 2 mg/kg of free ASO. Most of the material (>90 %) extravasated from the blood within the first 30 minutes followed by a short elimination phase (Tab. 3.2). C_{max} values of the free ASO (9090 ng ASO/ml) are much higher compared to the equally dosed liposomal group (3848 ng ASO/ml) indicating a much higher distribution of the liposomal ASO into the first compartment.

3.1.3 Quantitative organ distribution

Beside the pharmacokinetic of free and encapsulated Cy5.5-labeled ASO the biodistribution of this material into liver, spleen and kidney was followed. The quantitative organ distribution at different doses of nov038-Cy5.5-ASO 8 h and 24 h after injection is shown in Fig. 3.4.

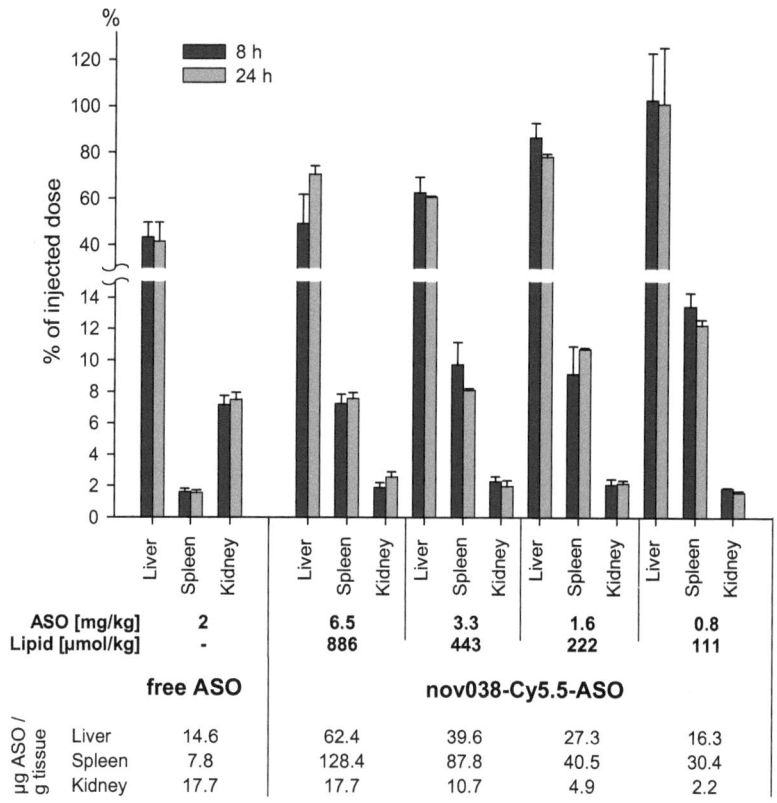

Fig. 3.4: **Organ distribution of free ASO and various doses of nov038-Cy5.5-ASO.** The uptake of labeled ASO into liver, spleen and kidney expressed as % of injected dose was determined after 8 h and 24 h. Data are represented as bars showing the mean [n = 3] ± SEM. The ASO deposition per gram tissue (after 8 h) is shown in table form in the lower part. Both, table and graph clearly show a dose-dependent uptake of nov038-Cy5.5-ASO into the primary organs liver and spleen. With increasing doses nov038 delivers increasing amounts of ASO into liver and spleen in a dose-dependent but not dose-linear fashion. Only the ASO uptake by the kidney seems to be dose-linear.

As expected, nov038 predominantly distributed into liver and spleen. A dose-dependent increase in liver and spleen uptake is visible in which the Cy5.5 signal increases in these primary organs with decreasing dose. Low lipid doses of nov038-Cy5.5-ASO (e.g. 111 µmol lipid/kg BW; ~4 µmol lipid/mouse) accumulated almost completely in the liver and spleen. However, high lipid doses (886 µmol/kg BW; ~32 µmol lipid/mouse) are taken up by the primary organs to an amount of only 50-60 % (16-19 µmol lipid). Liver and spleen uptake and binding sites were saturated with increasing lipid dose resulting in prolonged blood circulation times (see section 3.1.2) and accumulation of the material, for instance, in the kidney. The renal uptake of the Cy5.5-labeled ASO slightly increases with increasing dose of nov038-Cy5.5-ASO. The PK and BD of lower doses of nov038-Cy5.5-ASO was finished very early since no tremendous increase in liver, spleen and kidney uptake was seen after 24 h compared to 8 h. In contrast, nov038-Cy5.5-ASO injected at the highest dose (6.5 mg ASO/kg BW) showed an increase in liver uptake from 50 % to 70 % over the time because liposomes are still persistent in the blood circulation after 24 h and the distribution into the tissues was not totally completed at this measuring point.

Free, non-encapsulated ASO injected at a dose of 2 mg/kg accumulated in the liver, spleen and kidney with ~40 %, 1.6 % and 7.5 % of the injected dose, respectively (Fig. 3.4). Hence the uptake of the free ASO by the kidneys is 3 times higher compared to the liposomal groups (2-2.5 %). Most of the injected free ASO was renally cleared from the blood stream within the first 8 h and subsequently excreted by the urine. Free ASO disappeared from the blood circulation very rapidly (see section 3.1.2). Thus, no major difference in the BD was determined after 8 h and 24 h.

3.1.4 Microscopic distribution

The cellular distribution of Cy5.5-labeled ASO within different tissues was determined using epifluorescence (EFM) and confocal laser scanning microscopy (CLSM). At first, near infrared (NIR) fluorescence scans of tissue cryosections (liver, spleen and kidney) were recorded to achieve a more detailed tissue distribution (Fig. 3.5 A). Liver sections indicate a high uptake and a complete penetration of the massive organ by liposomal Cy5.5-labeled ASO whereas less intense Cy5.5 signals were determined in the liver section of free ASO, both injected at comparable doses. The distribution of liposomal ASO into spleen and kidney is less homogenous and depends on the tissue cells. In spleen sections only the red pulp (r) showed a Cy5.5 staining whereas the oval white pulp (w) region are non-stained. In kidney sections, extensive Cy5.5 staining was observed in the renal capsule and cortex (c) and was mainly located in the proximal tubules, but less Cy5.5 staining was observed in the renal pelvis (p). Mice treated with saline gave no NIR fluorescence signal in organ sections.

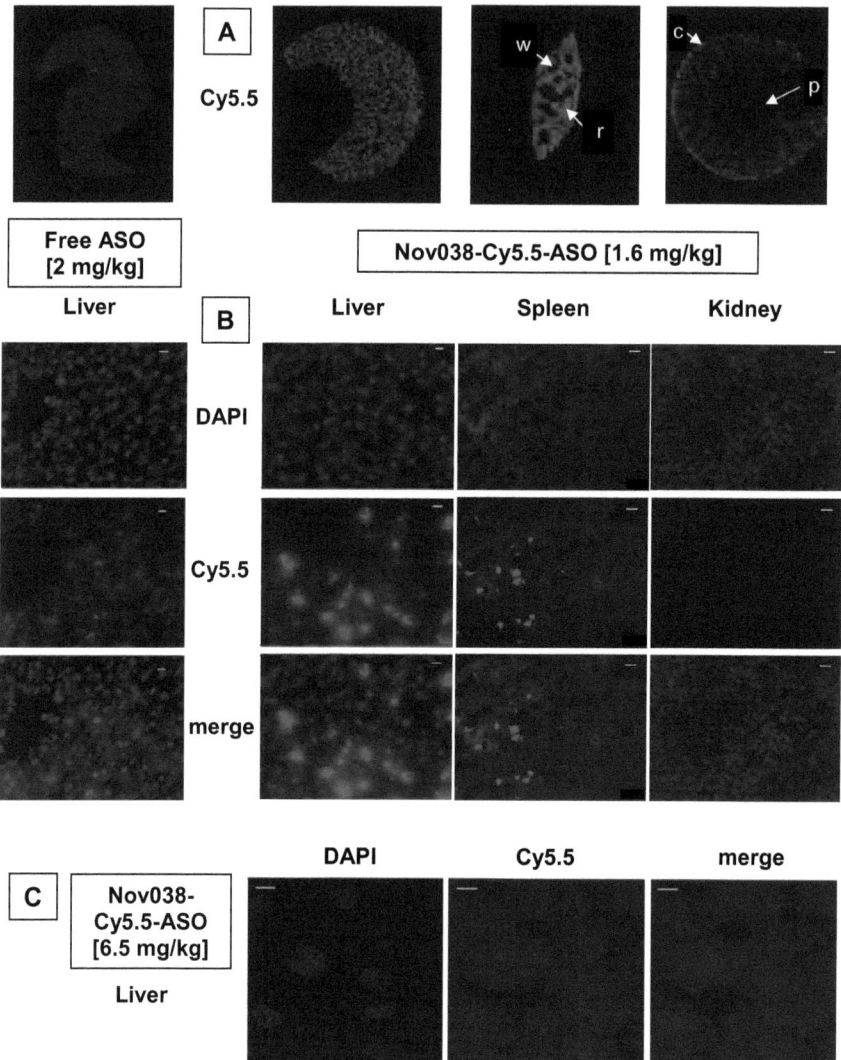

Fig. 3.5: **Microscopic tissue and cellular distribution of either free or formulated Cy5.5-ASO.** All images were taken 24 h after treatment. Organ cross sections from A) and B) were taken from mice treated with comparable doses of either free or formulated Cy5.5-labeled ASO [~2 mg/kg]. **A)** NIR fluorescence scans of tissue cryosections (liver, spleen and kidney). Red signals indicate Cy5.5 fluorescence in cross sections. c = cortex; p = renal pelvis; w = white pulp; r = red pulp. **B)** For comparative analysis fluorescence images were taken from sections of liver, spleen and kidney. Upper panel (blue): cells following DAPI staining. Mid-panel (red): near infrared fluorescence of Cy5.5. Lower panel (merge): overlay of DAPI and Cy5.5 images. White bar: 10 µm **C)** Confocal laser scanning microscopy (CLSM) of liver cross section treated with 6.5 mg/kg of nov038-Cy5.5-ASO indicating the uptake of labeled ASO into the cytoplasm. White bar: 7.5 µm

The EFM images following DAPI staining (nuclei), the near infrared images (Cy5.5) and the resulting overlay (merge) are presented in Fig. 3.5 B. For nov038, EFM confirms the highest Cy5.5 signal density in liver and spleen. The cellular distribution of nov038 within the liver and spleen sections appears in a more local and heterogeneous manner. In both cases red hot-spots are visible and, as well, a weak but broad staining of the respective parenchyma. Most likely, these Cy5.5-hot-spots represent tissue macrophages within the liver and splenic red pulp clearing the liposomes from circulation. Due to the less uptake of nov038-Cy5.5-ASO by the kidneys no signal could be determined in the renal pelvis by EFM. Fluorescence images of the free ASO group show a reduced uptake by the liver but the tissue staining is more homogenous compared to the formulated ASO at equal doses.

CLSM was conducted at liver sections of nov038-Cy5.5-ASO injected at a dose of 6.5 mg/kg. Using this high dose sample a clear uptake of the Cy5.5-labeled ASO into the cytoplasm and cell nucleus 24 h after injection could be shown (Fig. 3.5 C). The entire intracellular lumen as well as the nucleus of the hepatocytes is stained red. Again, no Cy5.5 signal could be determined in saline treated liver samples.

3.1.5 Determination of plasma AST/ALT levels and proinflammatory cytokines

At the end of the study [24 h after injection] plasma levels of liver enzymes AST and ALT were determined. Mean AST plasma levels (Tab 3.4) of mice treated with nov038-Cy5.5-ASO or free ASO were on average lower [66-98 U/l] in comparison to plasma levels in mice treated with saline [mean value of 100.5 U/l]. Mean ALT in the plasma of mice treated with saline or with nov038-Cy5.5-ASO or free ASO ranged between 95-149 U/l. AST and ALT levels in plasma of individual mice exhibited a high variability, especially saline treated mice. However, plasma AST and ALT levels were not elevated with respect to the control group.

To exclude a potential immunogenicity triggered by the lipid drug carrier and/or oligonucleotides a profile of proinflammatory cytokines (Il-1ß, Il-6, TNF) and interferon gamma (IFNγ) of different doses of nov038-Cy5.5-ASO was determined in plasma samples. No relevant elevation of cytokines or IFNγ was seen except for Il-6 (Fig. 3.6). Within the first hours [2-4 h post-injection] the elevation of the Il-6 plasma level by approx. 100 fold was seen at all tested doses of nov038-Cy5.5-ASO. After peaking at 4 h the Il-6 plasma level decreases and 24 h post-injection Il-6 reaches nearly baseline concentration. Saline data were obtained only after 24 h post-injection and a cytokine profile over time is missing.

Results

		AST			ALT		
		U/l	SD	%	U/l	SD	%
Saline		100.5	43.5	100.0	148.5	32.5	100.0
free ASO	2 mg/kg	66.0	8.0	65.7	95.0	36.2	64.0
nov038-Cy5.5-ASO	6.5 mg/kg	98.3	10.8	97.8	145.3	20.9	97.9
	3.3 mg/kg	81.3	16.4	80.9	95.0	28.6	64.0
	1.6 mg/kg	80.7	6.9	80.3	116.0	25.9	78.1
	0.8 mg/kg	68.0	10.7	67.7	105.7	33.5	71.2

Tab 3.4: Mean values [n = 3] of AST and ALT plasma levels [U/l], standard deviation [SD] and normalized expression [%] of values referred to saline treated group. Parameters were determined in blood plasma 24 h after single injection. A student's t-test for parametric data was used but no significant alterations were determined in groups of active treatment.

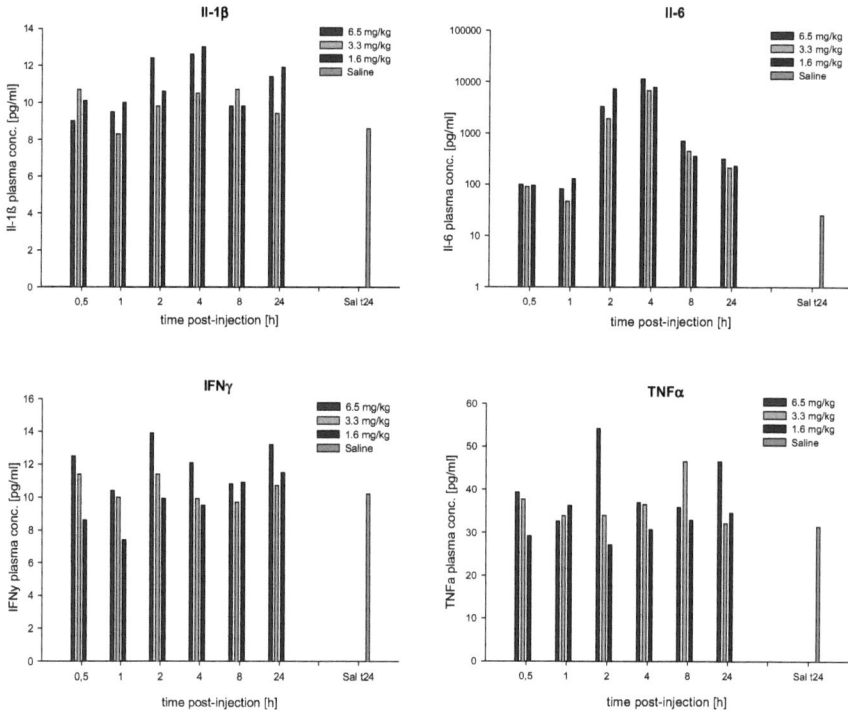

Fig. 3.6: Cytokine profile (Il-1ß, Il-6, TNFα and IFNγ) of nov038d213 collected within the first 24 h following a single injection. Cytokine concentrations [pg/ml] in plasma from one mouse per group were determined by ELISA in duplicates. Data are represented as bars showing the mean [n = 2] without error bars. Cytokines of a saline treated mouse determined 24 h after injection served as a control (Sal t24). No decisive alterations of cytokines were visible in the liposomal treated groups over the time except for Il-6 plasma levels which increased within the first 4 h after injection and subsequently decreased close to baseline after 24 h.

3.2 Pharmacodynamic of nov038-LT1-ASO

Following the PK/BD study the quantitative and qualitative influence of nov038 encapsulating a therapeutic active ASO on an *in vivo* mouse model was investigated. The pharmacodynamic study was performed in naïve BALB/c mice using an active ASO targeting a mouse liver mRNA (liver target 1, LT1). Within this study, the focus was set on the specific down-regulation (knockdown) of the hepatic LT1 mRNA and protein compared to a LT2 ASO encapsulated into nov038 which served as a non-LT1 control. Saline treated mice as well as free, non-encapsulated ASO (LT1 and scrambled) served as further controls.

Sample and study parameters for nov038-LT1 (nov038d145) and nov038-LT2 (nov038d144) are presented in Tab. 3.5. The particle size did not change considerably during the manufacturing, concentration, buffer exchange, and sterile filtration and was in the final product at 86 nm for nov038-LT1 and at 89 nm for nov038-LT2. The encapsulation efficiencies were ~55 % for both formulations. ASO concentrations based on the OD_{260} value were set to a final value of 1.25 mg/ml. The final lipid concentrations were 98 mM for both formulations resulting in a drug-to-lipid ratio of 12.8 µg ASO/µmol lipid.

	Saline	free ASO LT1 / LT2 / scr	nov038-ASO LT1			nov038-ASO LT2
Lot#	-	-	d145			d144
Avg. size [nm] / PI	-	-	86 / 0.17			89 / 0.22
ASO conc. [µg/ml]	-	1250	1250	625	125	1250
Lipid conc. [mM]	-	-	98	49	10	98
Drug-to-lipid ratio [µg/µmol]	-	-	12.8			12.8
Injection volume [µl]	200	each 200	200			200
ASO dose [mg/kg]	-	each 10	10	5	1	10
Lipid dose [µmol/kg]	-	-	784	392	78	784

Tab. 3.5: Nov038d145 and nov038d144 sample and study parameters. Both formulations were produced to equal average particle sizes, ASO and lipid concentrations. For injection purposes free ASO was diluted from stock solutions to a final concentration of 1.25 mg/ml. Male six weeks old BALB/c mice grouped to a number of five were treated twice a week for three weeks. For all samples the injection volume was 200 µl.

The treatment protocol was scheduled into five mice per group with two iv injections per week for three weeks at dosages as listed in Tab. 3.5. The quantification of LT1 mRNA (real-time PCR) and protein levels (Western blot) is shown in Fig. 3.7.

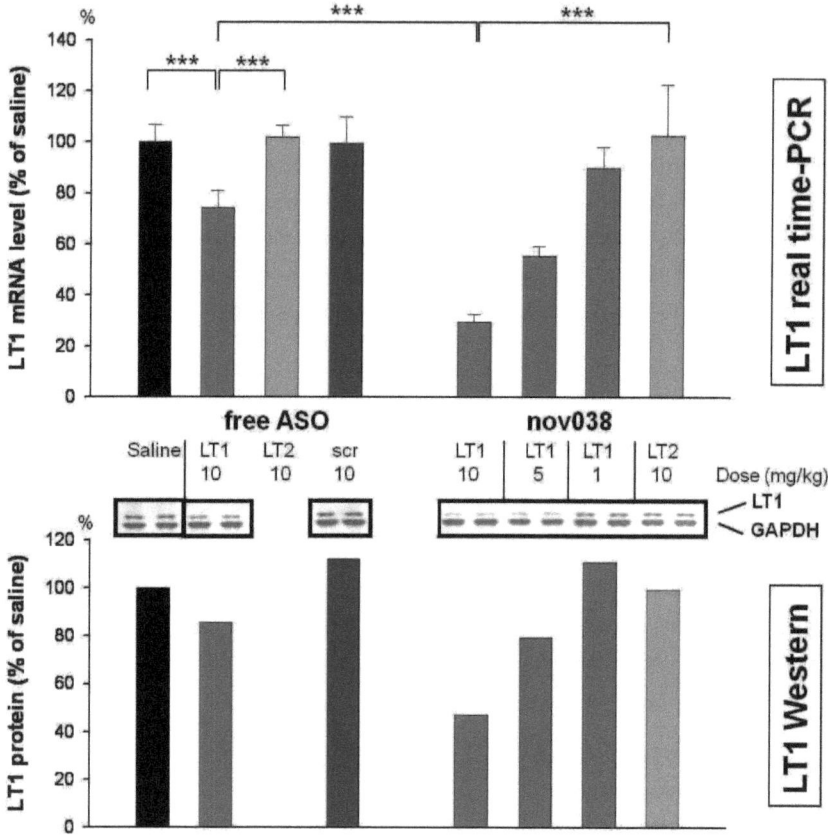

Fig 3.7: LT1 mRNA and protein analyses of mouse liver samples treated with free or encapsulated LT1 and control (LT2, scr) ASO. The upper graph shows the quantification of LT1 mRNA normalized to GAPDH mRNA levels and referred to the saline treated group. Data are represented as bars showing the mean [n = 5] ± SD. The lower graph illustrate the LT1 protein quantification from pooled liver samples [n = 5] plotted as duplicates. Average densitometrical values relative to the control values (saline) are plotted on bars. *** = $p < 0.001$

The present study demonstrates a dose-dependent down-regulation of the target mRNA level in the liver as determined by real time PCR. At a dosage of 10, 5 or 1 mg/kg of LT1-ASO encapsulated into nov038, the remaining mean LT1 mRNA expression was 30 %, 55 % or 90 %, respectively, relative to the saline control (Fig. 3.7; upper graph). Free LT1 ASO at a dosage of 10 mg/kg, showed a LT1 mRNA down-regulation of only 25 %. Control groups treated with scrambled or LT2-ASO at a dosage of 10 mg/kg injected either free or encapsulated into nov038 showed no significant down-regulation of the LT1 mRNA. A student's t-test analysis confirmed a strong significance for the down-regulation of the LT1 mRNA level by nov038-LT1 ($p < 0.001$) compared to free ASOs and nov038-LT2 at equal dosages of 10 mg/kg. Western Blot analysis and quantification of the LT1 protein from pooled liver samples [$n = 5$; plotted as duplicates] are illustrated in the lower graph of Fig. 3.7. The major result from western blot analysis was that treatment with LT1 ASO loaded into nov038 Smarticles led to a dose-dependent down regulation of the LT1 protein in the liver. At a dosage of 10 or 5 mg/kg liposomal LT1-ASO, the remaining mean LT1 protein expression was 43 % or 80 %, respectively, relative to the saline control. The expression of LT1 protein in the liver of mice treated with free LT1-ASO at a dosage of 10 mg/kg was reduced by 15 %. Free scrambled and LT2 ASO and liposomal LT2-ASO were used as controls for non-specific target down regulation. Mice treated with either oligonucleotide at a dosage of 10 mg/kg did not inhibit LT1 protein expression.

On both, the protein and the mRNA levels, only mice treated with the LT1 ASO loaded into nov038 exhibited a dose-dependent target down regulation. In contrast, analyses of liver samples of mice treated with free LT1 antisense at a concentration of 10 mg/kg demonstrate only slight LT1 reduction on protein and mRNA levels. The encapsulation of LT1-ASO into nov038 potentiated the target down-regulation by 3-fold on mRNA and 2-fold on protein level compared to the free LT1-ASO, both at a dosage of 10 mg/kg.

At the end of the study plasma AST and ALT levels were determined to investigate potential side-effects of the treatment after multiple dosing (Tab. 3.6). Mean ALT plasma levels of mice treated with nov038 with encapsulated ASO or free ASO were 31.0-53.4 U/l or 47.6-54.4 U/l, respectively, in comparison to plasma levels in mice treated with saline [mean value of 36.0 U/l].

All active groups showed no significant alterations of ALT plasma levels compared to the saline group except for nov038-LT1 injected at a dosage of 10 mg/kg. A student's t-test revealed a slight but significant increase of the ALT plasma level within this group. Mean AST levels in the plasma of mice treated with saline or with nov038-ASO or free ASO ranged from 82.0-96.2 U/l. Only mice of the group treated with nov038-LT1 at the lowest dosage showed an increased mean AST level of 154.2 U/l. However, a student's t-test did not confirm a significant increase within this group compared to the saline control.

		AST			ALT		
		U/l	SD	%	U/l	SD	%
free ASO	Saline	94.4	38.3	100.0	36,0	3,7	100,0
	LT1 10 mg/kg	82.0	19.0	86.9	47,6	16,8	132,2
	LT2 10 mg/kg	88,2	60,5	93,4	30,2	5,2	83,9
	scr 10 mg/kg	96.2	22.9	101.9	54,4	19,1	151,1
nov038-ASO	LT1 10 mg/kg	96.2	41.7	101.9	53,4*	15,8	148,3
	LT1 5 mg/kg	85.2	33.6	90.3	31,0	7,1	86,1
	LT1 1 mg/kg	154.2	98.9	163.3	44,8	24,6	124,4
	LT2 10 mg/kg	96.0	70.5	101.7	38,8	7,9	107,6

Tab. 3.6: Mean values [n = 5] of AST and ALT plasma levels [U/l], standard deviation [SD] and normalized expression [%] of values referred to saline treated group. Student's t-test for parametric data was used whereas the active treatment was compared with saline treatment. Only mice of the group treated with the highest dose of nov038-LT1 showed a slight but significant increase in plasma ALT levels (* = p < 0.05).

3.3 Proof-of-concept study using nov038-ApoB-siRNA

The previous study showed a significant knockdown of a liver target mRNA and protein (LT1) using nov038 with encapsulated LT1 antisense oligonucleotides. A second pharmacodynamic study aimed at the sequence specific down-regulation of a liver target using RNA interference (RNAi). An ApoB100 siRNA (ApoB I) was loaded into nov038 and tested in the liver "ApoB100 model" after systemic administration. At this point the control formulation nov038 with encapsulated scrambled (scr) siRNA was excluded from the study.

Batch production and study parameters for nov038-ApoB I (nov038d197) are presented in Tab. 3.7. Nov038 liposomes exhibited a mean particle sizes of 119 nm after concentration, buffer exchange, and sterile filtration. The encapsulation efficiency was ~50 % and less than 10 % of the total material was determined as non-encapsulated (outside) siRNA. Based on the OD_{260} value of the final product the liposomal suspension was diluted with PBS to a final siRNA concentration of 0.8 mg/ml. The final lipid concentration was 68 mM for nov038d197 resulting in a drug-to-lipid ratio of 11.8 µg siRNA/µmol lipid and thus comparable to the formulations nov038d144 and –d145 used in the previous study.

	Saline	nov038-ApoB I siRNA
Lot#	-	d197
Ave. size [nm] / PI	-	119 / 0.22
siRNA conc. [µg/ml]	-	800
Lipid conc. [mM]	-	68
Drug-to-lipid ratio [µg/µmol]	-	11.8
Injection volume [µl]	250	250
siRNA dose [mg/kg]	-	8
Lipid dose [µmol/kg]	-	680

Tab. 3.7: Nov038d197 sample and study parameters. Nov038 with encapsulated ApoB I siRNA was produced with an average particle size of 119 nm and was concentrated to a final siRNA and lipid concentration of 800 µg/ml and 68 mM, respectively. C57Bl/6 mice grouped to a number of five were treated every day for three days. For all samples the injection volume was 250 µl. Mice were sacrificed 24 h following the last injection.

The treatment protocol scheduled three iv injections on three consecutive days into five mice per treatment group at a dosage of 8 mg/kg of siRNA encapsulated into nov038. The mice were sacrificed 24 h following the last administration; liver samples were collected and prepared for quantification of apoB100 mRNA (Quantigene). Further, plasma was prepared from whole blood samples and was used for the quantification of ApoB100 protein (Western blot) and the determination of total cholesterol (Chol), HDL and LDL levels. The results are summarized in Fig. 3.8.

In contrast to the previous study using encapsulated ASO a triple dose of 8 mg ApoB I siRNA/kg BW encapsulated into nov038 did not lead to any decisive knockdown on ApoB100 mRNA or protein level (Fig. 3.8 A & B) and, further, no lowered total cholesterol or LDL levels were visible (Fig. 3.8 C). Mean plasma values from mice treated with nov038-ApoB I-siRNA seemed to be slightly increased and showed a high variance within the group, especially LDL values.

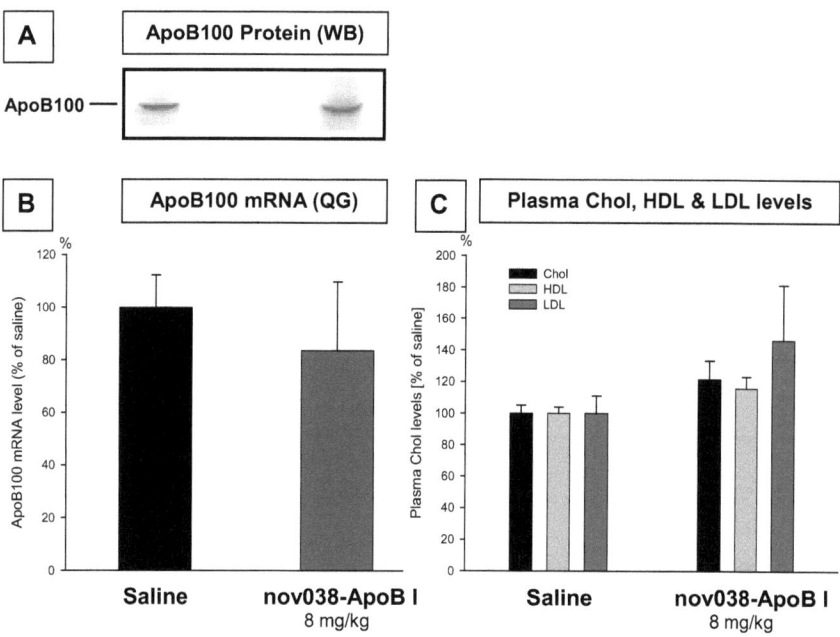

Fig. 3.8: **ApoB100 mRNA and protein as well as plasma cholesterol analyses. A)** Western Blot analysis of the ApoB100 protein in pooled plasma samples [n = 5]. **B)** Quantification of apoB100 mRNA within the liver using the Quantigene assay. ApoB100 mRNA levels were normalized to PPIB mRNA and referred to the saline control group. **C)** Total cholesterol (Chol), HDL and LDL levels were determined in plasma samples. Plasma levels of mice treated with nov038-siRNA were referred to the saline treated group. In summary, no significant down-regulation of the ApoB100 mRNA or protein and LDL-cholesterol plasma levels could be determined after treatment with nov038 encapsulating ApoB I siRNA. Data from B) and C) are represented as bars showing the mean [n = 5] ± SD.

3.4 *In vitro* transfection of primary mouse hepatocytes (PMHs) using nov038 loaded with either ASO or siRNA

PK/BD analyses of nov038 loaded with ASO molecules reveal a lipid dose-dependent blood circulation and distribution into the primary organs liver and spleen. Assuming that the PK/BD of liposomes only depends on carrier parameters (size, charge, dose etc.) the encapsulated type of oligonucleotide has no influence on the extracellular distribution. However, regarding the down-regulation of a target mRNA *in vivo*, totally different results were obtained when using nov038 loaded with either ASO or siRNA molecules. Whereas nov038-LT1-ASO reduced the liver mRNA and protein levels in a dose-dependent manner no decisive down-regulation was visible using RNAi. In a one-step-back approach the transfection efficiency of nov038 loaded with either ASO or siRNA molecules was investigated *in vitro* on primary mouse hepatocytes (PMHs). Particular attention was paid to the endosomal escape of labeled oligonucleotides after intracellular delivery by nov038. It is known that ASO and siRNA molecules show different escape pathways, which might be responsible for the different knockdown effects.

3.4.1 Transfection of PMHs with ASO and siRNA molecules targeting apoB100 mRNA using the cationic transfectant jetPEI-Gal

At first, different oligonucleotides targeting the apoB100 mRNA were tested on freshly isolated and cultivated hepatocytes to compare the silencing potency of each oligonucleotide. For comparative analyses an ApoB100 ASO containing locked-nucleic acid (LNA) modifications was taken from Swayze and co-workers, 2007.[139] The ApoB I siRNA from previous studies was further modified by a 5'-phosphorylation at the antisense strand and was tested head-to-head *in vitro* on PMHs. Scrambled Control (scr) ASO and scr siRNA served as controls and all oligonucleotides were testes at concentrations ranging from 1...10 nM/well. A buffer (0.1x PBS) treated group served as a control. Mean normalized apoB100 mRNA values refer to the buffer treated control group and are shown in Fig. 3.9.

Fig. 3.9 clearly shows the down-regulation of apoB100 mRNA using target specific siRNA or ASO molecules. Both oligonucleotides, ApoB I 5'P siRNA and LNA-ApoB ASO reduced the apoB100 mRNA level by ~80 % compared to the buffer treated group at a concentration of 1 nM. An improvement regarding the knockdown of apoB100 mRNA by using the ApoB I 5'P siRNA is clearly visible (KD ~95 %) as the ApoB I siRNA showed a KD of ~72 % at equal concentrations. The corresponding scrambled control oligonucleotides showed no significant reduction of the target mRNA level. For further comparative analyses both, the LNA-modified ApoB-ASO as well as the ApoB I 5'P siRNA were used.

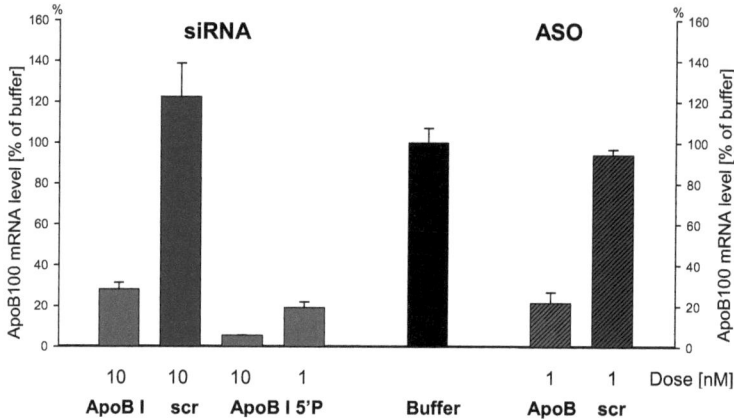

Fig. 3.9: Comparative analysis of different types of ASO and siRNA molecules targeting the apoB100 mRNA. Target mRNA levels were normalized to PPIB mRNA and referred to the buffer control group. Data are represented as bars showing the mean [n = 3] ± SD. Groups treated with complexed siRNA molecules are shown on the left side in filled bars and ASO treated groups are shown on the right side in shaded bars. ApoB100 oligonucleotides treated groups are shown in red whereas the respective scrambled control groups are highlighted in blue color.

3.4.2 Transfection of PMHs with ASO and siRNA molecules targeting apoB100 mRNA encapsulated into nov038

Both oligonucleotides, the LNA-modified ApoB-ASO as well as the ApoB I 5'P siRNA, were encapsulated into nov038. Also, the corresponding scrambled control oligonucleotides were loaded into nov038 and served as controls. Batch production parameters are presented in Tab. 3.8. Nov038 liposomes loaded with siRNA molecules exhibited mean particle sizes of ~150 nm. During the encapsulation of ASO molecules liposomes were formed with mean particle sizes of 164 nm and 203 nm for nov038-LNA-ApoB ASO and nov038-scr-ASO, respectively. For comparative analysis nov038 liposomes were tested at different doses with oligonucleotide concentrations of up to 1000 nM on PMHs.

Results from the apoB100 mRNA Quantigene assay are depicted in Fig. 3.10. Up to a concentration of 1000 nM ApoB I 5'P or scrambled siRNA loaded into nov038 no significant alteration of the apoB100 mRNA level was visible. In contrast, a decisive down-regulation of ~60 % compared to buffer treated group was determined at a concentration of 100 nM LNA-ApoB ASO loaded into nov038. Further increasing of the LNA-ApoB ASO dose (up to 1000 nM) did not result in higher down-regulation of the apoB100 mRNA level. Even low concentration of LNA-ApoB ASO (10 nM) showed a significant KD of ~25 % and an IC_{50}

value of ~80 nM was calculated for nov038-LNA-ApoB ASO. As expected, a dose of up to 1000 nM of scrambled ASO encapsulated into nov038 did not lead to an alteration of the target mRNA level.

	nov038-siRNA		nov038-ASO	
	ApoB I 5'P	scrambled	LNA-ApoB	scrambled
Lot#	d231	d232	d233	d234
Ave. size [nm] / PI	154 / 0.20	152 / 0.21	164 / 0.17	203 / 0.21
Oligo conc. [µg/ml]	165	165	73	79
Oligo conc. [µM]	11	11	11	11
Lipid conc. [mM]	9.7	9.7	3.2	3.3
Oligo concentration tested on cells [nM]	10...1000	1000	10...1000	1000

Tab. 3.8: Sample and study parameters for nov038 loaded with different types of oligonucleotides. Both formulations encapsulating either ASO or siRNA were produced to nearly equal average particle sizes. Final liposomal suspensions were pre-diluted to a concentration of 11 µM oligonucleotide using PBS (Gibco). Liposomes were tested at the highest concentrations of 1000 nM of encapsulated siRNA and ASO. For further doses tested on hepatocytes, the liposomes were diluted appropriately.

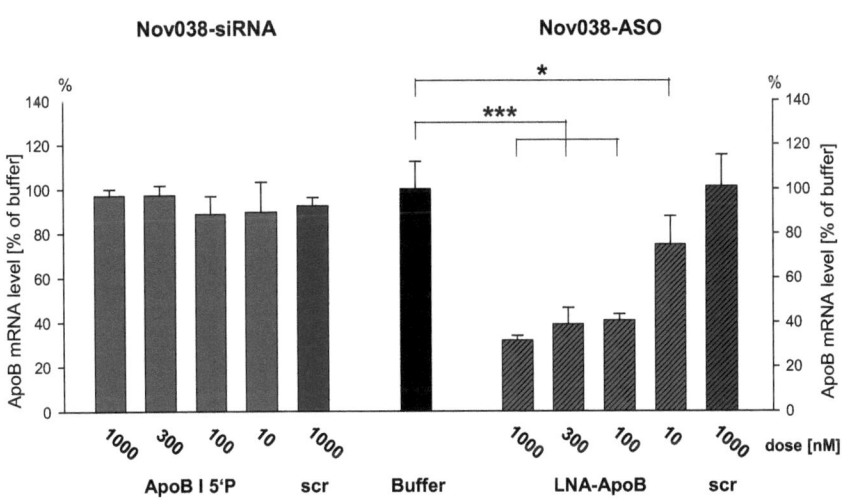

Fig. 3.10: Comparative analysis of nov038 loaded with either siRNA (left side) or ASO (right side, shaded bars) molecules targeting the apoB100 mRNA. Only nov038 loaded with an LNA-modified ApoB-ASO significantly reduced the apoB100 mRNA. Nov038 with encapsulated siRNA had no effect on target down-regulation. Data are represented as bars showing the mean [n = 3] ± SD. * = $p < 0.05$; *** = $p < 0.001$

The current in vitro study confirmed the findings discovered in vivo in that a clear down-regulation of a specific liver mRNA and protein was shown using an antisense oligonucleotide encapsulated into nov038 whereas no decisive comparable effect was visible using RNAi. As aforementioned ASO molecules are able to cross the cell and endosomal membrane, presumably via receptor-mediated endocytosis or channels [2], whereas naked or unformulated siRNA molecules are not transported across the membranes. The (intact) delivery of siRNA molecules into the target cells using nov038 cannot be seen from the previous studies. Studies with formulated labeled oligonucleotides should reveal their uptake and subcellular distribution. This may give evidence for the oligonucleotide translocation pathway and/or possible barriers on the way to the cytoplasm

3.4.3 Uptake of nov038 loaded with Cy5.5-labeled ASO or siRNA by PMHs

Due to different delivery efficiencies of nov038 loaded with either ASO or siRNA molecules the uptake and disposition of nov038 loaded with fluorescently labeled oligonucleotides on hepatocytes was followed. Therefore, suspensions of nov038 loaded with either Cy5.5-labeled siRNA or Cy5.5-labeled ASO were used for transfection. Sample parameters for nov038d235 and nov038d222 are presented in Tab. 3.9.

	Nov038	
	Scr-Cy5.5 siRNA	Scr-Cy5.5 ASO
Lot#	d235	d222
Ave. size [nm] / PI	133 / 0.20	162 / 0.12
Oligo conc. [µg/ml]	15.8	35.3
Oligo-Cy5.5 conc. [µg/ml]	15.8 (100%)	7.1 (20%)
Oligo-Cy5.5 conc. [µM]	1.1	1.1
Lipid conc. [mM]	2.0	6.5
Oligo-Cy5.5 concentration tested on cells [nM]	100	100

Tab. 3.9: Sample and study parameters for nov038 loaded with ASO or siRNA molecules both labeled with Cy5.5. Nov038d235 was loaded with a pure scr-Cy5.5 siRNA whereas for the production of nov038d222 a mixture (4:1; w/w) of unlabeled scr-ASO and scr-Cy5.5 ASO was used. Both formulations were pre-diluted to a final oligonucleotide concentration of 1.1 µM using PBS (Gibco).

Liposomes of nov038 loaded with a pure scr-Cy5.5-labeled siRNA had a mean particle size of 133 nm whereas a mixture of CD40-ASO and scr-Cy5.5-labeled ASO (4:1; w/w) was used for loading into nov038 forming liposomes with a mean particle size of 162 nm. Both formulations were tested at Cy5.5-labeled oligonucleotide concentration of 100 nM.

The hepatocytes were washed with PBS 4 h following the transfection step to remove non-transfected or membrane-bound liposomes from the media and were prepared for confocal laser scanning microscopy (CLSM) 20 h following the intermediate washing step.

nov038-Cy5.5-siRNA   nov038-Cy5.5-ASO

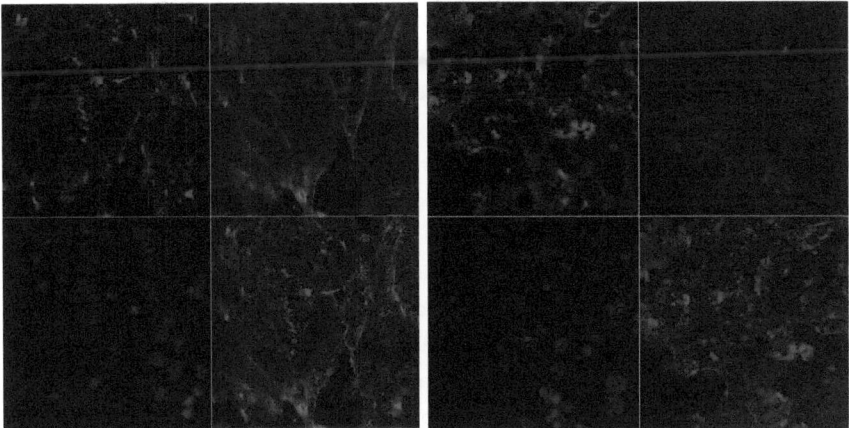

Fig. 3.11: CLSM images of hepatocytes treated with either nov038-Cy5.5-siRNA (left side) or nov038-Cy5.5-ASO (right side). Cy5.5 fluorescence signals are shown in red. Hepatocytes were further counter-stained with DAPI (nuclei in blue) and TRITC-conjugated phalloidin staining F-actin molecules (green). A merged image for both treatments is shown in the right lower corner, respectively. Both formulations were taken up by the cells and both show a dominant endosomal localization of the label, especially perinuclear.

The hepatocytes were counterstained with DAPI (nuclei in blue) and TRITC-conjugated phalloidin staining F-actin molecules (green). Cy5.5 fluorescence signals from labeled oligonucleotides are shown in red (Fig. 3.11). Both formulations were taken up by the cells as both showed a dominant endosomal localization of the label. This spotty disposition appeared predominantly perinuclear. Nov038 showed a roughly equal uptake for both, siRNA and ASO, whereas a clear disposition of siRNA and ASO molecules within the cytoplasm or nucleus, respectively, could not be visualized.

Along the cellular delivery mediated by nov038 the distribution patterns of labeled oligonucleotides give hints for a breaking-off within the endosome. While the single-stranded ASO may be able to translocate into the cytoplasm / nucleus after degradation of nov038 during endosomal / lysosomal maturation the double-stranded siRNA will be probably degraded within the destructive lysosomal environment.

3.5 Rational design of Smarticles and application for PMHs

The previous studies revealed the delivery of antisense oligonucleotides to the target cells using the liposomal formulation nov038. The ability of nov038 to deliver ASO molecules to hepatocytes *in vitro* and *in vivo* was shown by quantitative mRNA and protein analyses. However, using siRNA molecules loaded into nov038 no decisive down-regulation of the target mRNA or protein levels was detectable.

In contrast to single-stranded antisense oligonucleotides, the double-stranded siRNA molecules are not able to cross the membrane because of their high hydrophilicity and the absence of specific transferring receptors. Hence, the active and intact delivery of siRNA across the plasma membrane is an absolute condition to achieve biological effects. Therefore novel liposomal formulations were designed to be both, stable and highly fusogenic. The rational design of highly fusogenic liposomes, as described in the chapter "Introduction – 1.6 Lipid Shape Theory", included a highly fusogenic state at the isoelectric point of the lipid mixtures at a pH of 5 - 6. This would lead to a fusion of the liposomes with the endosomal membrane and the liberation of the cargo into the cytoplasm. Further these novel Smarticles formulations exhibit stable states at low pH for loading of oligonucleotides and at physiological pH for storage and application purposes.

As outlined before the key parameter of the new model, κ_{MIN}, describes the geometry of the ion-free lipid membrane around the isoelectric point and is thought to be predictive for the fusogenicity of the lipid assembly. To test whether κ_{MIN} predicts the ability of amphoteric liposomes to transfect siRNA molecules into cells, esp. hepatocytes, several formulations were designed and prepared for transfection of primary mouse cells. Therefore, chemically distinct amphoteric liposomes were loaded with ApoB I 5'P siRNA. Lipid compositions as well as κ_{MIN} values of the respective formulations used in this study are summarized in the Appendix (section 8.2 & 8.3). Usually, the liposomes were used for transfection of hepatocytes at a siRNA dose range of 10...1000 nM. The reduction of apoB100 mRNA levels were determined three days after transfection and IC_{50} values, the concentration of siRNA needed to reduce the target mRNA by 50 %, were deduced for all formulations and plotted against the respective κ_{MIN} value (Fig. 3.12). IC_{50} values of samples showing no reduction up to a siRNA concentration of 1000 nM were set to 1000 nM. A tight correlation between efficacy and κ_{MIN} was observed upon apoB100 mRNA targeting in primary mouse hepatocytes. The transfection of hepatocytes was limited by κ_{MIN} and low values of about 0.13...0.16 were required for optimized cellular transfection. No appreciable cellular transfection was observed for κ_{MIN} values >0.22. Plots for nov038 and nov729 are highlighted in red (Fig. 3.12). The κ_{MIN} value for nov038 was calculated to 0.35. As shown before, no mRNA reduction was detected up to a concentration of 1000 nM. In contrast, nov729 (κ_{MIN} = 0.145) efficiently transfected hepatocytes with an IC_{50} value of ~22 nM.

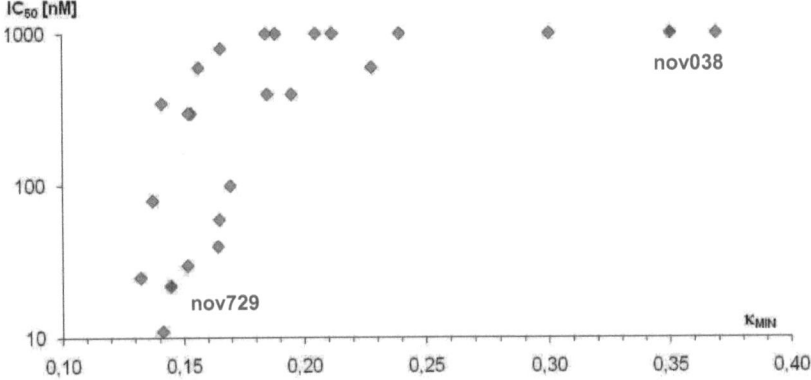

Fig. 3.12: IC_{50} vs. K_{MIN} plot. Chemical distinct amphoteric liposomes were loaded with ApoB I 5'P siRNA and IC_{50} values were calculated from quantified apoB100 mRNA levels of a distinct dose range (10…1000 nM) tested on hepatocytes. A clear dependency between low IC_{50} and K_{MIN} values is visible. A small K_{MIN} value is necessary for a high efficacy of formulations to deliver siRNA. Plots for nov038 and nov729 are shown in red, exemplarily.

A closer examination of the lipid composition, calculations thereof and the fusion behavior of both formulations, nov038 and nov729, is given in Tab. 3.10 and Fig. 3.13. Whereas Amphoter II systems (comprising charge-reversible cationic lipids) allow for cation:anion ratios (ratio C:A) of ≥1 (nov038 = 1) Amphoter I class liposomes (with permanent cationic lipids) have to have a ratio C:A of <1 (nov729 = 0.67) to maintain an overall anionic surface charge at physiological pH. Thus, nov038 comprises a portion of 50 % anionic CHEMS of total charged lipids and nov729 contains 60 % of anionic DMGS of total charged lipids.

nov038 [II]				[Amphoter system]	nov729 [I]		
MoChol	Chems	POPC	DOPE	composition	DODAP	DMGS	Chol
20	20	15	45	mol %	24	36	40
			60	% neutral lipids			40
	40			% charged lipids		60	
	1			Ratio C:A		0,67	
	50			% anionic of charged lipids		60	
	0,350 @ 6.0			K_{min} @ IP		0,145 @ 5.6	

Tab. 3.10: Differentiated examination of the lipid composition of nov038 and nov729. Ratio C:A defines the molar quotient of cationic to anionic lipids. K_{min} and the isoelectric point (IP) of the distinct lipid mixture were calculated from the algorithm of the dynamic shape theory.

The κ_{min} value for each distinct formulation is calculated at the isoelectric point (IP) and represents a key fusion parameter. Variations of the anionic lipid portion and the influence on the fusion properties are shown in Fig. 3.13. The lipid composition of nov038 shows high κ values (up to 0.42) over the entire pH range (*in silico* data Fig. 3.13 A, calculated and drawn up from the algorithm of the dynamic shape theory [129]). A valley is calculated at pH 6.0 with a κ value of 0.35 (refers to κ_{min}). Increasing the amount of anionic lipid slightly decreases the κ values and shifts the IP into the acidic pH range. The decrease of κ values with 100 % of anionic CHEMS of total charged lipids (= 40 % CHEMS and 60 % of neutral lipids with a ratio C:A=0) and the increase in fusion propensity could be proven with experimental fusion data (Fig. 3.13 B). Homofusion of these liposomes, which is defined as the fusion between particles of the same lipid composition, was shown in the pH range of 2.5 - 4.5 (dusky pink) whereas the lipid mixture of nov038 showed no fusion propensity over the entire pH range. However, lipid mixtures with only anionic lipids cannot be used for efficient encapsulation of nucleic acids. Further decreasing the amount of CHEMS leads to higher κ values with no improved fusogenicity.

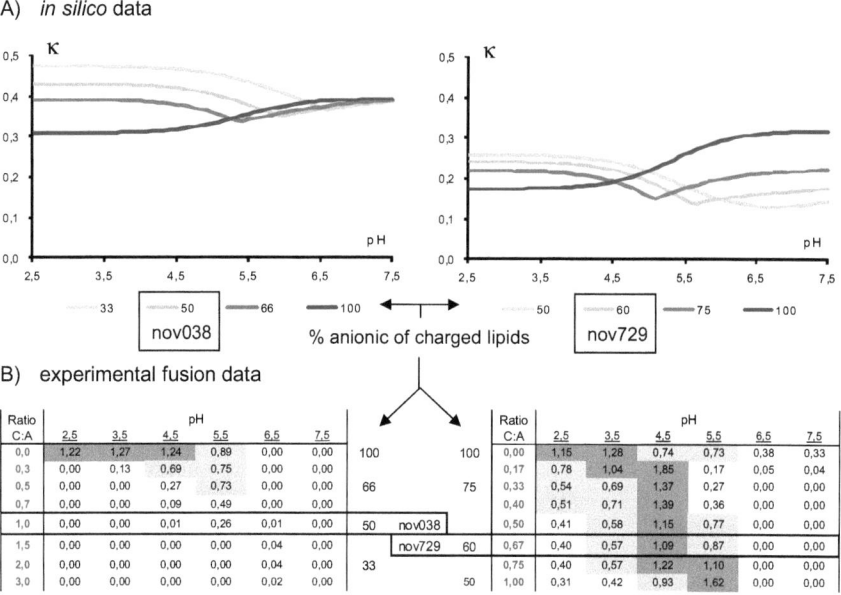

Fig. 3.13: *In silico* calculation of κ values over a pH range of 2.5 – 7.5 (A) for nov038 (left panel) and nov729 (right panel) and the conformation of κ-dependent fusion obtained by experimental fusion data (B). Both, *in silico* and experimental fusion data show a landscape of the respective formulation with various amounts of anionic lipid. The cation:anion molar ratio (ratio C:A) was changed from 0 to 3 for MoChol:CHEMS (B, left panel, Amphoter II system) and from 0 to 1 for DODAP:DMGS (B, right panel, Amphoter I system). Experimental fusion data were kindly provided by Evgenios Siepi and details for experimental setup can be found in [127]. Intensity of fusion expressed as FRET-signals E590/530: 0…0.5 = no color; 0.5…1 = pink; 1…2 = dusky pink.

In contrast, the lipid mixtures of nov729 and variations thereof exhibit substantially lowered κ value compared to nov038. Nov729 with 60 % of anionic DMGS of total charged lipids shows a bi-phasic curve with higher κ values in the pH range of 2.5 – 4.5 and at physiological pH indicating non-fusogenic, stable phases. A valley is calculated at pH 5.6 with a κ value of 0.145 (κ_{min}) where the highest fusion tendency is predicted. Experimental fusion data (Fig. 3.13 B) reveal the highest homofusion for nov729 between pH 4.5 – 5.5 (colored dusky pink) and no fusion at strong acidic and physiological pH. Again, increasing the amount of anionic lipid promotes fusion in the strong acidic pH range whereas less amounts of DMGS shifts the κ_{min} values of the formulation to an IP of ~6. This was further confirmed by experimental fusion data. Here, the height of the respective κ_{min} values were not affected by varying the amount of anionic lipids, only the IP and thus the pH range with the highest tendency of fusion.

3.6 Delivery of oligonucleotides using the fusogenic nov729

Within the previous trials 25 chemical distinct amphoteric liposomes were screened with respect to their efficacy to deliver siRNA molecules to hepatocytes *in vitro*. In summary, efficient down-regulation of apoB100 mRNA was achieved only with small κ_{MIN} value. One of these fusogenic candidates, called nov729, was selected for further analyses *in vitro* and *in vivo*. Based on previous considerations nov729 should also be able to deliver ASO molecules. In the following study both, ASO molecules and siRNAs loaded into nov729 were tested head-to-head on primary mouse hepatocytes.

3.6.1 Transfection of PMHs with nov729 encapsulating ASOs or siRNAs

Comparable to previous studies ApoB I 5'P siRNA was encapsulated into nov729 as well as the LNA-modified ApoB-ASO and their corresponding scrambled control oligonucleotides. Sample and study parameters are presented in Tab. 3.11. The production of nov729d017 and –d018 led to the formation of liposomes with equal mean particle sizes of 99 nm for both formulations. In contrast, ASO-loaded liposomes exhibited a somewhat larger mean particle size. For all batches of nov729 the encapsulation efficiency was ~65 %. Samples were tested at concentrations ranging from 10…100 nM.

	nov729-siRNA			nov729-ASO	
	ApoB I 5'P	scr	scr-Cy5.5	LNA-ApoB	scr
Lot#	d017	d018	---	---	---
Ave. size [nm] / PI	99 / 0.12	99 / 0.10	114 / 0.20	127 / 0.08	115 / 0.05
Oligo conc. [µg/ml]	16.5	16.5	17.0	7.3	7.9
Oligo conc. [µM]	1.1	1.1	1.1	1.1	1.1
Lipid conc. [mM]	1.0	1.0	1.0	0.6	0.6
Oligo concentration tested [nM]	10…100	100	100	10…100	100

Tab. 3.11: Sample and study parameters for nov729 loaded with apoB100 mRNA targeting siRNA and ASO molecules and scrambled controls. Both siRNA formulations were produced to equal average particle sizes (99 nm), siRNA and lipid concentrations. Liposomes loaded with ASO molecules were somewhat larger in particle size compared to the nov729-siRNA formulations. All samples were pre-diluted to an oligonucleotide concentration of 1.1 µM to achieve a final concentration of 100 nM on cells. The apoB mRNA targeting batches were further diluted to gain final oligonucleotide concentrations of 10 nM and 30 nM.

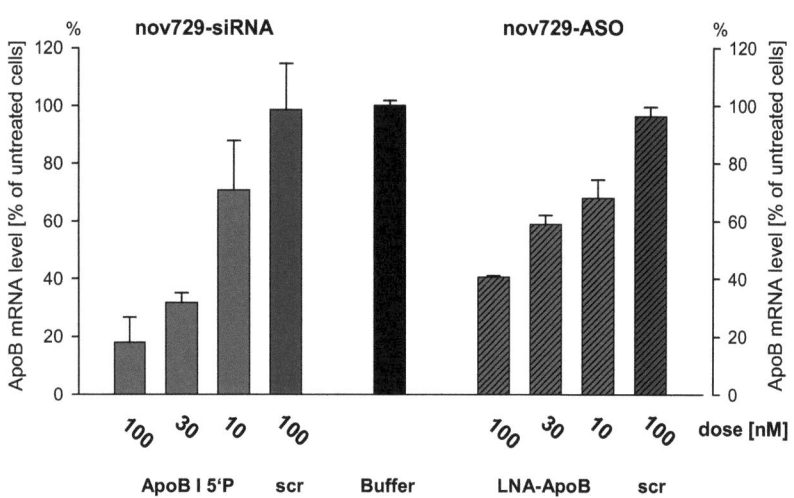

Fig. 3.14: Transfection of PMHs using nov729 loaded with apoB100 mRNA targeting oligonucleotides. This fusogenic formulation mediated an apoB100 mRNA reduction using RNAi and antisense technology in a dose-dependent manner. The target mRNA downregulation is sequence-specific as the encapsulated scrambled oligonucleotides showed no effect. Data are represented as bars showing the mean [n = 3] ± SD

Mean normalized apoB100 mRNA values referred to the buffer treated control group are shown in Fig. 3.14. In stark contrast to nov038 (Fig. 3.10) the fusogenic formulation nov729 mediated a dose-dependent down-regulation of the apoB100 mRNA with encapsulated ApoB I 5'P siRNA. A dose of 100 nM, 30 nM and 10 nM of the target siRNA led to a mean mRNA reduction of >80 %, 68 % and 29 %, respectively, compared to the buffer treated group. Further, the observed reduction of the apoB100 mRNA was sequence-specific as the treatment with scrambled siRNA encapsulated into nov729 had no effect upon the apoB100 mRNA. Moreover, nov729 mediated a dose-dependent knockdown of the apoB mRNA with ASO molecules. The ASO-triggered KD seems to be somewhat weaker than the RNAi-mediated downregulation. A dose of 100 nM, 30 nM and 10 nM of the LNA-ApoB ASO led to a mean mRNA reduction of ~60 %, 40 % and 32 %, respectively.

3.6.2 Pharmacodynamic of nov729 loaded with ApoB I 5'P siRNA *in vivo*

With respect to the latter *in vitro* study indicating a sequence-specific RNAi effect nov729 was subsequently tested within the "ApoB100 model" *in vivo*. Therefore the ApoB I 5'P and a scrambled control siRNA were encapsulated into nov729, called nov729d004 and nov729d005, respectively. Sample and study parameters for both formulations are presented in Tab. 3.12. Particles were produced with equal mean particle sizes of ~106 nm. Encapsulation efficiency was ~70 % for both formulations and non-encapsulated (outside) siRNA was not detectable. The final lipid and siRNA concentration were ~22 mM and 0.5 mg/ml, respectively, for both formulations resulting in a drug-to-lipid ratio of ~23 µg siRNA/µmol lipid. Mice grouped to a number of four to five were injected twice (day 1 and day 3). The quantification of liver apoB100 mRNA and plasma cholesterol, HDL and LDL levels are shown in Fig. 3.15.

In contrast to the *in vitro* studies no conclusive down-regulation of the apoB100 mRNA in the liver and plasma protein (data not shown) could be determined *in vivo* using nov729-ApoB I 5'P (Fig. 3.15; A). A dosage of 8 mg ApoB I 5'P siRNA/kg BW slightly decreased the apoB100 mRNA level but was not significant to the saline treated group. However, a significant down-regulation (* = $p < 0.05$) by nov729-ApoB I 5'P siRNA was determined compared to the nov729-scr-siRNA treated group, both at a dosage of 8 mg/kg. At dosages of 4 mg siRNA/kg BW for both, active and control formulation, a slight increase of the apo100 mRNA level was visible.

	Saline	Nov729-siRNA	
		ApoB I 5'P	scrambled
Lot#	-	d004	d005
Ave. size [nm] / PI	-	106 / 0.12	107 / 0.14
siRNA conc. [µg/ml]	-	500	500
Lipid conc. [mM]	-	22.0	21.5
Drug-to-lipid ratio [µg/µmol]	-	22.7	23.2
Injection volume [µl]	200	400 / 200	400 / 200
siRNA dose [mg/kg]	-	8 / 4	8 / 4
Lipid dose [µmol/kg]	-	352 / 176	344 / 172

Tab. 3.12: Production and study parameters for nov729d004 and nov729d005. Both formulations were produced to equal average particle sizes, siRNA and lipid concentrations. For injection purposes the liposomal batches were diluted to a final siRNA concentration of 500 µg/ml. Mice grouped to a number of four to five were injected twice (day 1 and day 3). Liver and plasma were sampled 24 h following the last injection.

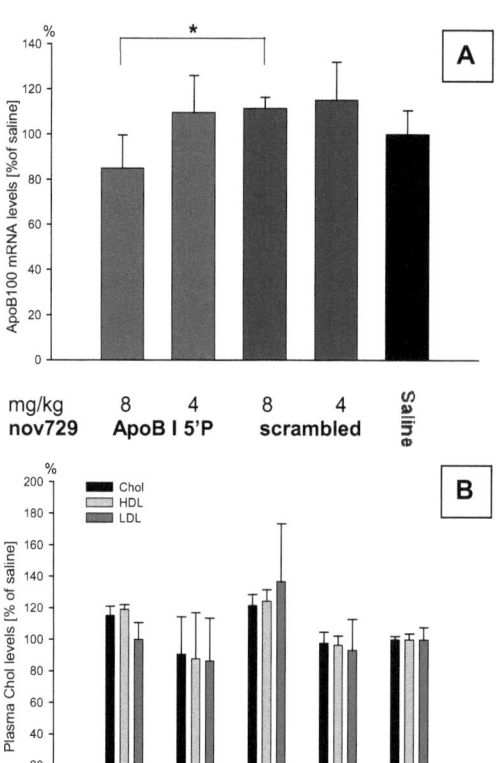

Fig. 3.15: ApoB100 mRNA and plasma cholesterol analyses after treatment with nov729 encapsulating ApoB I 5'P or scrambled control siRNA.
A) Quantification of apoB100 mRNA within the liver using the Quantigene assay. The apoB100 mRNA levels were normalized to PPIB mRNA and referred to the saline control group.
B) Total cholesterol (Chol), HDL and LDL levels were determined in plasma samples. Plasma levels of mice treated with different doses of nov729-siRNA were referred to the saline treated group. Data from A) and B) are represented as bars showing the mean [n = 4-5] ± SD. In summary, no substantial down-regulation of the apoB100 mRNA or protein (data not shown) and LDL-cholesterol plasma levels could be determined after treatment with nov729 encapsulating ApoB I 5'P siRNA compared to the saline treated group. * = $p < 0.05$

SiRNA dosages of 8 mg/kg slightly increased the plasma cholesterol levels (Fig. 3.15 B). Mean values of plasma cholesterol levels are slightly reduced in the ApoB I 5'P siRNA treated group (8 mg/kg) compared to the respective scrambled control group, especially the mean plasma LDL values (but with high variance within this group). Total cholesterol, HDL and LDL plasma levels of groups treated with either 4 mg/kg of nov729-ApoB I 5'P siRNA or nov729-scr-siRNA showed no significant alterations compared to the saline treated group. Also, the ApoB I 5'P siRNA treated groups (at 4 mg/kg) exhibited a high biological variance. In summary, only slight reductions of apoB100 mRNA and plasma LDL levels were visible after treatment with nov729-ApoB I 5'P siRNA compared to the scrambled control group, both at dosages of 8 mg/kg.

To give evidence about potential toxicities using nov729 *in vivo* plasma AST and ALT levels were determined at the end of the study (Tab. 3.13). Generally, no significant alterations could be detected within the liposomal treated groups compared to the saline group. Mean AST levels were slightly reduced, except for the group nov729 treated with a dosage of 4 mg ApoB I 5'P siRNA/kg BW whereas all liposomal treated groups showed slightly elevated mean ALT levels compared to the saline group. Usually, increased mean ALT values involved high variances within the groups.

		AST			ALT		
		U/l	SD	%	U/l	SD	%
	Saline	83.8	15.7	100.0	37.3	5.8	100.0
Nov729	ApoB I 5'P; 8 mg/kg	73.0	15.0	87.2	40.3	13.9	108.1
	ApoB I 5'P; 4 mg/kg	86.4	31.8	103.2	42.0	13.9	112.8
	scrambled; 8 mg/kg	73.3	17.0	87.5	49.3	17.4	132.2
	scrambled; 4 mg/kg	83.0	18.1	99.1	41.0	6.0	110.1

Tab. 3.13: **Mean values [*n* = 4-5] of AST and ALT plasma levels [U/l], standard deviation [SD] and normalized expression [%] of values referred to saline treated group.** Student's *t*-test for parametric data was used but no significant alterations were calculated. No significant elevations of AST and ALT plasma levels were determined indicating nov729 as a safe formulation.

3.6.3 Transfection of PMHs with nov729 in the presence of mouse serum

Using nov729 with encapsulated ApoB I 5'P siRNA *in vivo* no efficacy regarding target mRNA and protein down-regulation or physiological effect (lowering of cholesterol and LDL plasma levels) could be shown. *In vitro*, a clear sequence-specific knockdown of the apoB100 mRNA was shown using nov729 with encapsulated siRNA for transfection of hepatocytes. *In vitro* studies on HeLa cells indicated an inhibition of the transfection efficiency of nov729 by lipoproteins (see Appendix; Fig. 8.2). In this context, nov729 loaded with ApoB I 5'P siRNA (nov729d017; Tab. 3.11) was tested on hepatocytes in the presence of complete mouse serum at concentrations of 100...1000 nM. Cells were supplemented with complete mouse serum to a final concentration of 10 % immediately before adding the liposomes. Transfection efficiencies in terms of apoB100 mRNA downregulation in the presence of mouse serum are shown in Fig. 3.16 A.

In the absence of mouse serum nov729 loaded with ApoB I 5'P siRNA mediated a dose-dependent knockdown of the apoB100 mRNA level. A dose of 100 nM and 300 nM led to an mRNA reduction by 75 % and 95 %, respectively, compared to the buffer treated group. The latter dose highlights the maximum knockdown of the apoB100 mRNA level since a dose of 1000 nM ApoB I 5'P siRNA did not lead to a further mRNA reduction. In contrast in the presence of 10 % complete mouse serum the effect is completely diminished while using a dose of 100 nM of nov729-ApoB I 5'P siRNA. By increasing the dose up to 300 nM and 1000 nM the mRNA reduction effect is nearly restored. The inhibition by mouse serum is thus dose-dependent and is diminished by high concentrations of liposomes which in turn titrate the inhibiting serum components. The presence of mouse serum had no effect on the apoB100 mRNA level as shown in both buffer treated groups, with and without mouse serum. Further, the inhibition of cell transfection by mouse serum was demonstrated by CLSM (Fig. 3.16 B). Nov729 was loaded with a Cy5.5-labeled siRNA resulting in the formation of liposomes with a mean particle size of 114 nm (Tab. 3.11). A siRNA concentration of 100 nM was tested on hepatocytes and cells were prepared for CLSM 20 h following the intermediate washing step. First of all, CLSM images clearly demonstrate the intracellular delivery of siRNA molecules mediated by nov729 (Fig. 3.16 B; left and central image). The Cy5.5-labeled siRNA is located within the cytoplasm and only little vesicular distribution of the label is visible within the cell. By adding mouse serum the quantitative amount of cytoplasmic Cy5.5-labeled siRNA decreased indicating a strong inhibition of the transfection efficiency of nov729 (right image).

Finally, nov729 failed to enable an RNAi-mediated target down-regulation *in vivo*, presumably due to an inhibition by lipoproteins. Further development, e.g. the design of novel lipid head groups and liposomal compositions is necessary to overcome these barriers.

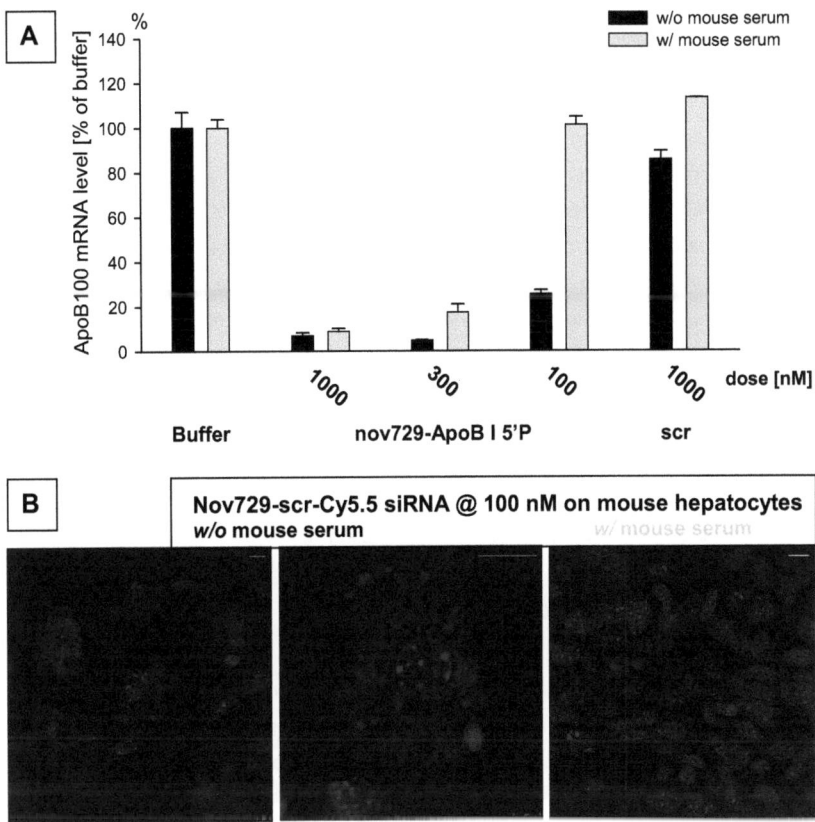

Fig. 3.16: **Transfection of hepatocytes using nov729 with encapsulated siRNA and inhibition by mouse serum.** A) Reduction of apoB100 mRNA level after treatment with nov729-ApoB I 5'P siRNA in the presence or absence of mouse serum. At an ApoB I 5'P siRNA dose of 100 nM the target KD is totally abrogated when co-administered with complete mouse serum. The inhibition is dose-dependent; a siRNA dose of 300 nM led to an apoB100 mRNA reduction of 80 % compared to the buffer treated group in the presence of mouse serum. Doses as high as 1000 nM led to the complete titration of mouse serum and no differences were visible regarding the reduction of apoB100 mRNA. Data are represented as bars showing the mean [n = 3] ± SD. B) CLSM images of nov729-transfected hepatocytes. Cells were washed with PBS 4 h following the transfection step to remove liposomes from the media. The hepatocytes were prepared for CLSM 20 h following the washing step. The uptake of encapsulated Cy5.5-labeled siRNA was followed and fluorescence signals thereof are shown in red. Hepatocytes were further counter-stained with DAPI (nuclei in blue). Merged images for the treatment w/o serum (two magnifications) and with serum are shown. Nov729 clearly shows the delivery of siRNA into the cytoplasm (left and central image) but in the presence of mouse serum the transfection efficiency is strongly diminished (right image). White bar: 10 μm

4. Discussion

This work sets out to develop and describe amphoteric liposomes for the delivery of oligonucleotides *in vitro* and *in vivo*. Besides functionality and mode of action, these drug carriers have to be characterized in terms of pharmacokinetic (PK) and biodistribution (BD). The present work initially emphasizes with a dose-dependent PK and BD of nov038, a Smarticles formulation with a known ability to transfect macrophages and dendritic cells *in vivo* [86], and basically describes its pharmacological behavior. The PK/BD of liposomes is lipid-dose dependent and thus nov038 was formulated to a rather low final drug-to-lipid ratio to achieve high lipid doses (~1000 µmol/kg) at therapeutic interesting ASO doses of ~10 mg/kg. Besides the lipid dose, the particle size decisively influences the clearance and tissue distribution of liposomes.[88,111,146] With an average particle size of 174 nm and a narrow size distribution pattern nov038 liposomes are able to penetrate into the liver parenchyma (<150 nm) but will be further detected by cells of the MPS. Additionally, conclusions should be drawn for application of nov038 in hepatocytes-targeting strategies.

4.1 PK of nov038 is non-linear and depends on lipid dose

The administered lipid dose has a crucial impact on circulation times and tissue distribution of the liposomes after intravenous application. Increasing doses (<1...>1500 µmol lipid/kg BW) substantially prolong the blood half-life of liposomes and compartments such as resident macrophages in liver and spleen can be saturated.[88,119] This leads to elevated blood levels and distribution of liposomes into different tissue and organ compartments.[119]

A follow-up dose-dependent study describes the PK of Smarticles formulation nov038 in whole blood and its quantitative organ distribution over time. The PK profile of nov038 depicted in Fig. 3.3 and Tab. 3.2 as well as the PK parameter (Tab. 3.3) show a dose-dependent elimination of the liposomal formulation from the blood stream. Liposomes of nov038 can be described by a non-linear pharmacokinetic in which a dose proportionality of the initial and terminal half-life ($t_{1/2}$ [h]), C_{max}, AUC and CL_{tot} is not observable. With respect to their properties and conditions these PK parameters altered in a dose-related, but not dose-linear fashion. With increasing lipid doses of nov038 in a stepwise manner the initial and terminal $t_{1/2}$ rose also. Linear PK models, however, are characterized by constant $t_{1/2}$ independent of the injected dose. Also, in a linear PK the total body clearance (CL_{tot}) is constant and independent of the dose or blood level. The non-linear decrease of CL_{tot} as well as the non-linear increase of C_{max} and AUC by doubling the lipid doses (Tab. 3.3) argues for a non-linear PK of nov038 in a two-compartment model with two parallel elimination pathways. The two-compartment model is further supported by the fast uptake of nov038 into liver and spleen (first compartment) shown in Fig. 3.1 and Tab. 3.2. The most common reason for non-linear PK is the existence of saturable elimination, distribution or binding mechanisms. Typically, for liposomes, two kinetic compartments are described in the literature, wherein one is saturable and the other is non-saturable.[90,94,102] During the fast distribution phase nov038 liposomes will be taken up from the blood circulation by phagocytic cells in liver and the splenic red pulp. After saturation of this first distribution site the liposomes extravasate into the liver parenchyma. The more lipid dose was injected the more particles spilled over the primary distribution sites which further leads to longer circulation time in the blood. Low doses of nov038 were rapidly cleared from the blood stream and only one half-life could be calculated from mono-exponential regression fits. Data points from early distribution phases are missing and the description of the early PK behavior of low dose nov038 is thus rather incomplete. However, high doses of nov038 clearly show a two-phased clearance of the liposomes with a relatively short and fast primary distribution and a dose-dependent prolonged elimination phase.

4.2 Nov038 distributes into saturable compartments

Nov038 liposomes rapidly (Fig. 3.1) and predominantly distribute into liver and spleen and a dose-dependent increase in hepatic and splenic uptake is shown (Fig. 3.4 table in the lower part). However, the relative Cy5.5 signals (expressed as % of injected dose) increase in the primary organs with decreasing dose (Fig. 3.4 graph). The BD is finished very early since no significant increase in liver and spleen uptake was seen at the later time point except for the highest dose of nov038. Here, the uptake into liver, spleen and kidney after 24 h is somewhat higher maybe due to the substantial prolonged circulation time at this dose. Scans of the dissected organs further reveal less or no signal in the lungs, kidneys and heart (Fig. 3.2). The majority of NIR-signals detected in the kidney, however, are caused by non-encapsulated Cy5.5-labeled ASO. A constant amount of 13 % of non-encapsulated, outside ASO was administered per injection at each dose (Tab. 3.1) resulting in a 2 % ASO recovery in the kidneys. Also, with prolonged blood circulation small amounts of ASOs may liberate from the liposomes leading to an elevated renal ASO uptake and clearance.

Close to these findings previous internal studies (report nov-025-2005) suggest rapid initial kinetics together with high overall adsorption at liver and spleen for the amphoter II liposomes. Studies were conducted in rats but are comparable to those in mice in terms of lipid doses. Single iv doses of approx. 100 µmol lipid/kg BW of a radiolabeled nov038 suspension resulted in an initial and terminal $t_{1/2}$ of 2 min and 60 min, respectively. In addition, further decreasing doses (e.g. 20 µmol lipid/kg BW) did not substantially alter the PK/BD of nov038 in rats indicating the requirement of a minimum lipid dose to overcome the first compartment. As shown in this study lipid doses higher than 111 µmol/kg BW led to prolonged circulation times. Conclusively, the lower the lipid dose the more lipid material is relatively distributed to the first compartment. A fixed portion of the injected lipid is bound by the first compartment until a saturation of the binding sites is reached.

For liposomes containing DOPE:CHEMS a similar fast blood clearance was described for comparatively low lipid doses (<100 µmol/kg BW) in that the pH-sensitive liposomes were almost completely eliminated from the bloodstream within 0.5 h.[147] In contrast, the terminal $t_{1/2}$ was dramatically enhanced by introducing PEGylated lipids ($t_{1/2}$ = 11.1 h) and a substantial percentage of liposomes (8.5 %) remained in the blood after 24 h. Also, the main distribution site for pH-sensitive liposomes, either PEGylated or not, was the liver and spleen.

As a major advantage, the present study indicates a long and stable circulation for the amphoteric formulation nov038 without the incorporation of PEGylated lipids. Increasing lipid doses of nov038 saturate the first distribution sites (liver and spleen) and gradually increase the blood residence time and bioavailability. Thus, residual particles of nov038 may distribute from the blood into peripheral compartments (e.g. sites of inflammation or cancer tissue).

4.3 Free, non-encapsulated (naked) ASO shows a rapid kinetic

For comparative analysis a group of mice was treated with free, non-encapsulated ASO at a dose of 2 mg/kg. Fig. 3.3 and Tab. 3.2 clearly show a rapid clearance of the free ASO from the blood stream. At nearly equal doses the terminal $t_{1/2}$ of free ASO is comparable to that of the encapsulated ASO. However, C_{max} values as well as AUC and CL_{tot} indicate a PK of the free ASO which is different from the liposomal PK. A lower AUC and higher CL_{tot} argue for a fast and massive elimination of the free ASO, presumably by the kidneys.

These findings are in total agreement with PK/BD data published by ISIS Pharmaceuticals in a rat study.[148] All tested phosphorothiolated (PS) ASO molecules, either 2'-MOE modified or not, showed comparable PK profiles characterized by a rapid initial distribution phase (initial $t_{1/2}$ of 0.2 - 0.3 h) followed by a much slower elimination phase (terminal $t_{1/2}$ of 1 - 3 h). Geary and co-workers reported the recovery of an equivalent ASO in liver>kidney>spleen with high uptake of the ASO molecules in the kidney cortex and efficient elimination in the urine. The recovery of the Cy5.5-labeled ASO in the urine is missing in this work but NIR-scans of a cryosected kidney reveal the highest concentration of the Cy5.5-labeled ASO within the kidney cortex, too (see Appendix, Fig. 8.1).

At therapeutic doses PS-modified ASO molecules also show a dose-dependent pharmacokinetic and tissue distribution.[149] Phillips and co-workers reported a biphasic blood kinetic of a 20mer PS-modified ASO characterized by a rapid initial clearance followed by a prolonged circulation with highest accumulation in kidney, liver and spleen. The distribution to high accumulating tissues was saturable and resulted in non-linear pharmacokinetics. For the kinetics of ASO molecules at least two compartments have to be assumed which are, however, different from the liposomal compartments.

The fact, that the liposomal entrapment of ASO molecules improves their PK profile (e.g. increased blood $t_{1/2}$, reduced CL_{tot}) and BD into liver and spleen, is widely accepted.[150] Thus, liposomes broaden the range of therapeutic application for ASOs and/or reduce therapeutic relevant doses in organs far from the kidneys, mainly liver and spleen.

4.4 Microscopy reveals uptake of nov038 by the liver parenchyma

A more detailed look towards the microscopic distribution reveals a broad and intense staining in NIR scans of liver sections (Fig. 3.5 A). In contrast, spleen sections showed a more localized staining in which the Cy5.5-label was found only in the red pulp. This region of the spleen parenchyma is well supplied with blood vessels and numerous macrophages are located in this area. Former internal FACS and *in vivo* knockdown studies revealed the uptake of nov038 by splenic macrophages and dendritic cells (preferentially by CD11b low or negative expressing subpopulations thereof; report nov-006-2009).[86] Oval, non-stained regions show the splenic white pulp, secondary lymphoid nodules (Malpighian bodies) which are characterized by a dense population of T- and B-lymphocytes. Thus, the uptake of nov038 by splenic lymphocytes is negligible.

In kidney sections (Fig. 3.5 A), Cy5.5 staining was observed in the renal capsule and cortex and was mainly located in the proximal tubules, but less Cy5.5 staining was observed in the renal pelvis (see also Appendix Fig. 8.1). The renal clearance of blood components by glomerular filtration strongly depends on their molecular size and charge (reviewed in [151]). Molecules with a MW of >50 kDa (~6 nm in diameter) will be retained by the glomerular basement membrane and cationic substances will be preferably cleared than anionic or neutral ones. Since liposomes of an average particle size of ~174 nm pass by the kidney staining results from free, non-encapsulated ASO (MW: 6.5-7.5 kDa; Tab. 3.1).[9]

Fluorescence images clearly show the distribution of nov038 into liver and spleen (Fig. 3.5 B). Intense Cy5.5 signals (red hot-spots) are most likely caused by delivery into tissue macrophages. However, this result needs to be confirmed with immunohistochemistry using macrophage specific antibodies (e.g. against CD68). Beyond the selective staining of cells of the MPS a broad staining of the liver and spleen parenchyma is visible. Confocal microscopy of selected liver parenchyma regions shows a distinct uptake of the material into hepatocytes after 24 h (Fig. 3.5 C). For CLSM liver sections of mice treated with the highest dose of nov038-Cy5.5-ASO were used to clearly illustrate the delivery of the formulated ASO molecules into the liver parenchyma. The entire intracellular lumen of the hepatocytes is stained red and an effective distribution of ASO molecules into the cytoplasm or nucleus is a prerequisite for the subsequent antisense-dependent mRNA degradation.

Images of free Cy5.5-ASO show a reduced uptake by the liver compared to the formulated ASO at equal doses. Further, the distribution of the free ASO in the liver is more homogenously and the uptake into cells of the MPS is negligible (Fig. 3.5 B).

4.5 High lipid doses of nov038 are non-toxic

Aspartate aminotransferase (AST, ASAT) and alanine aminotransferase (ALT, ALAT) are mainly present in the liver, but also in smaller amounts in the kidneys, heart, muscles, and pancreas. AST and ALT blood levels are rapidly raised in conditions that affect the heart and liver, amongst others extensive damages from toxins and drugs. Normal levels of AST and ALT in mice range from 80-400 U/l and 35-200 U/l, respectively, and depend on the mouse strain and environment (personnel communication Jonas Füner, Preclinics, Potsdam). Mean values of AST and ALT plasma levels [U/l] of mice treated with different doses of nov038 varied in the normal range and were not elevated with respect to the control group (Tab. 3.4). Plasma AST and ALT levels do thus not suggest liver damage caused by any single doses of the lipid drug carrier or ASO molecules. High lipid doses of nov038 were tolerated very well even after frequent dosing, as shown in Tab. 3.6.

In addition, a cytokine profile of nov038d213 was generated to clarify the potential immunogenic reactions upon the treatment with formulated oligonucleotides (Fig. 3.6). Proinflammatory cytokines (e.g. interleukin (II)-1ß, II-6, II12 and TNFα) are released from macrophages upon contact to antigens to trigger a subsequent immune response. Concomitantly, IFNγ is secreted by T-cells to stimulate the maturation of B-lymphocytes.

Compared to the control group no substantial and immunological relevant alterations in the cytokine levels at any of the time points or in dependency of the injected dose were determined except for Il-6. Plasma levels elevated within the first hours by approx. 100-fold at all tested doses and afterwards decreased to nearly baseline concentration. This curve progression is independent from the injected lipid dose meaning either that a small amount of lipid is necessary to induce the short elevation of Il-6. A lipid amount higher than the "activation" dose leads to no further cytokine induction. The Il-6 increase (up to a conc.$_{Il6}$ of 10 ng/ml) is not expected to cause significant immune stimulation and is nearly two orders of magnitude lower than those observed upon stimulation with lipopolysaccharides (LPS; 25 µg intraperitoneally injected) in mice (conc.$_{Il6}$ >500 ng/ml).[152] Mice from the treatment group with the highest dosing of nov038d213 received a single endotoxin/LPS injection of 1.25 EU (0.125 – 0.25 ng, determined by a commercial limulus amebocyte lysate (LAL)-test) which is not sufficient to induce a significant immunological response. A cytokine profile of saline-treated mice over time is missing but would be a benefit to interpret daytime variations in cytokine levels or elevations due to injection injuries caused by needles.

4.6 Nov038 delivers ASO but not siRNA molecules

The PK/BD study revealed a massive distribution of high doses of nov038 into the liver and microscopy of liver sections showed the uptake of the formulated Cy5.5-labeled ASO into liver parenchyma cells (hepatocytes). In a following pharmacodynamic (PD) study a therapeutic active ASO molecule targeting the liver LT1 mRNA was formulated and, comparable to the former PK/BD study, nov038 liposomes were prepared with a low drug-lipid-ratio to achieve ASO doses of 10 mg/kg at high lipid doses *in vivo* (Tab. 3.5). Further, nov038d144 and nov038d145 exhibited a much smaller average particle size (< 90 nm) to decrease the recognition by cells of the MPS. Both, high lipid dose and small particle size, improve the delivery of liposomes to the liver parenchyma.[88]

The formulation of LT1-ASO potentiated the antisense-dependent degradation of the LT1 mRNA by a factor 2-3 (Fig. 3.7). Six doses of 10 mg/kg of formulated LT1-ASO significantly reduced the LT1 mRNA by 70 % compared to the saline treated group whereas equal doses of free LT1-ASO led to an mRNA reduction of only 30 %. Control groups including free or liposomal LT2 or scrambled ASO showed no effect on the LT1 mRNA and protein expression levels. The gene silencing potency of the 2'MOE-modified LT1-ASO was shown by ISIS Pharmaceuticals (Carlsbad, CA, USA).[153] Similar results using free, non-encapsulated ASO were found by Swayze and co-workers.[139] Six doses á 18 mg/kg of a 2nd generation 2'MOE-modified ASO targeting the apoB100 mRNA led to a reduction of 50 % which is comparable to the findings in this work.

In this work, LT1 was used as an exemplary model target to evaluate the potency of nov038 delivering antisense agents to hepatocytes. LT1 is expressed in liver parenchyma cells as well as in liver macrophages (Kupffer cells)[154] and one might speculate that the elevated LT1 knockdown is due to an increased uptake by the phagocytic cells. However, nov038 liposomes exhibited a small particle size with a narrow size distribution and studies using nov038 with encapsulated ASO molecules against the hepatic diacylglycerol O-acyltransferase 2 (DGAT 2) revealed the nov038-mediated and antisense-dependent mRNA degradation in liver tissues (report nov-012-2006).

It is thought that, regarding sequence-dependent mRNA degradation, RNA interference is much more potent than the antisense approach but definitely needs the assistance by delivery agents to overcome the hurdles of blood circulation and cellular uptake. With respect to the successful ASO-mediated hepatic target mRNA knockdown in mice nov038 was used to deliver siRNA molecules to the liver parenchyma. Regarding a stringent vector development, PK/BD, efficacy and safety data for nov038 with encapsulated ASO molecules give a sufficient rationale for the subsequent testing of formulated siRNAs *in vivo*. However, the treatment with nov038-ApoB siRNA did not yield in any significant

Discussion

knockdown of the target mRNA or protein *in vivo* (Fig. 3.8 A and B). Further, no reduction of the total cholesterol (Chol) or LDL plasma level could be observed (Fig. 3.8 C) which would be a consequence of a knockdown of the scaffolding protein ApoB100 in the liver (see Tab. 1.2). In this study, nov038 exhibited a somewhat larger average particle size (119 nm) compared to the previous studies but is still preferred for an uptake by the liver parenchyma. A low drug-to-lipid-ratio with resulting high lipid doses should be sufficient to guarantee a considerable distribution into the liver parenchyma, as shown in the PK/BD and former PD study of nov038 and treatment and dosing schedule are comparable to other ApoB-siRNA approaches listed in Tab. 1.2. Soutschek and co-workers originally designed and validated the apoB100 targeting siRNA [133] and the sequence-specific silencing potency of ApoB I siRNA was also shown *in vitro*, on primary mouse hepatocytes (PMHs), in this work (Fig. 3.9). A 5'-phosphorylation at the anti-sense strand of ApoB I siRNA further improved the silencing potency. The 5'-phosphate is considered to be a prerequisite for the stability and slicing fidelity within the active RISC complex and acts as a key determinant for the strand selection.[155]

In vitro transfection studies on PMHs with nov038 confirmed the *in vivo* findings. Here, the knockdown of apoB100 mRNA triggered either by ASO or siRNA molecules loaded into nov038 was investigated. A highly potent LNA-modified ApoB100 antisense (taken from [139]) exhibited a sequence-specific knockdown efficiency comparable to the ApoB I 5'P siRNA when transfected with *in vivo*-jetPEI™-Gal (Fig. 3.9). However, only the ApoB-ASO mediated a reduction of the apoB100 mRNA when loaded into and transfected with nov038 whereas formulated siRNA doses of up to 1000 nM had no effect on the apoB100 mRNA level (Fig. 3.10). Both formulations were comparable in particle size and oligonucleotide concentration. The lipid concentration for nov038 with encapsulated siRNA is three times higher on PMHs than for the antisense formulations. The lower drug-to-lipid-ratio for nov038-siRNA results from the chosen process parameters and the particles of nov038d231 and d232 seem to be more multilamellar vesicles (MLV) or with a lower siRNA payload per particle resulting in a higher particle number. However, the difference in lipid amounts does not explain the vast difference between both oligonucleotide approaches. Further, subcellular localisation studies reveal an endosomal uptake and trapping of Cy5.5-labeled ASO and siRNA molecules encapsulated into nov038 (Fig. 3.11). CLSM images of *in vitro* transfected PMHs indicate a roughly equal uptake for both, formulated ASO and siRNA, and, seemingly, the spotty endosomal vesicles are located in the perinuclear region. Still, a more intense cytosolic staining is visible for nov038-ASO than for the corresponding siRNA formulation assuming the release of the single-stranded antisense from the endosome.

After endocytotic uptake across the cell membrane liposomes disintegrate due to a decreasing pH (e.g. acid-induced hydrolysis of phospholipids) [156] and influx of lipases (e.g.

degradation of liposomal lipids through (lyso-)phosphollipases A and C) [157] during endosome maturation.[158] At this stage single-stranded ASO molecules will be released from the liposomes and will escape from the endolysosomal compartment by using transporting proteins ("oligoportin") as reviewed by Bennett and Swayze, 2010.[2] Further, it could be shown that the cellular uptake of ASOs is an active and/or energy-dependent mechanism including endocytotic or pinocytotic pathways [55] or the transport through anion channels.[159] The cell entry of naked ASOs is time and temperature dependent and, in vitro, only 1-2 % of ASOs are taken up.[160] The delivery of free, non-encapsulated ASO molecules, their appearance in the cytoplasm/nucleoplasm and effectiveness in vivo is documented by numerous publications and the correlation of in vitro and in vivo delivered naked ASO was recently shown by Stein and coworkers (called "gymnosis").[57] However, in vitro doses of gymnotically delivered LNA-modified ApoB-ASOs are comparably high (IC_{50} >2.5 µM) whereas nov038 mediated an uptake of an equal antisense with 25-fold higher efficiency (IC_{50} <100 nM; Fig. 3.10). It can be assumed that nov038 liposomes are actively taken up by the cells leading to a concentration within endosomal vesicles and a higher cytoplasmic release of the loaded ASO molecules across the endosomal membrane. Moreover, the endosomal uptake of liposomes, their disintegration during endosomal/lysosomal maturation and translocation of ASO molecules seems to be a saturable process because increasing doses of nov038-ApoB ASO did not lead to a substantially elevated gene silencing.

In contrast, double-stranded siRNAs are not able to cross the membrane due to their high hydrophilicity and the absence of specific transferring receptors. Under acidic pH conditions, e.g. in maturating endosomes or lysosomes (pH <5), siRNAs are thought to dissociate or disintegrate.[161] The loss of siRNA integrity accompanies with a profound reduction in RNAi efficiency and intact siRNA duplexes are superior to single strands of siRNA by several orders of magnitude.[162] Hence, the active and intact delivery of siRNA across the plasma and endosomal membrane is an absolute condition to achieve biological effects. Therefore liposomal delivery systems were designed to guarantee I) a stable loading, storage and transport of siRNAs in biological fluids and II) safe and efficient membrane crossing of siRNAs into the cytoplasm by avoiding endolysosomal compartmentalization.

4.7 A rational design of fusogenic liposomes enables the effective delivery of siRNAs on PMHs

In this work a rational approach for the design of novel, fusogenic amphoteric liposomes was utilized. The algorithm which was used for a prediction of fusogenic liposomal mixtures bases on the dynamic lipid shape theory allowing for the molecular shape of neutral and ionizable cationic and anionic lipids, their binding of counterions in dependency of the environmental pH, the formation of an ion-free interlipid salt bridges and the resulting lipid

Discussion

membrane phase transition. Siepi and co-workers give a detailed mechanistic insight into the process of lipid fusion and its relationship to counterion binding.[127,129] Here, the correlation between fusogenicity of liposomal mixtures (expressed by the fusion determinant κ_{min}) loaded with siRNAs and their potential to transfect HeLa cells or the macrophage cell line RAW264.7 was observed.

In this work the evaluation of potent transfectants on PMHs based on the molecular analysis of RNAi-mediated reduction of the apoB100 mRNA expression level (Appendix, Tab. 8.2). Plotted data of the respective κ_{min} and IC_{50} values provide a link between the fusion determinant κ_{min} and effective siRNA delivery on PMHs (Fig. 3.12). Low values of κ_{min} were required for a potent siRNA delivery indicated by IC_{50} values of <30 nM. With gradually increasing κ_{min} values the transfection efficiency of these liposomal mixtures was substantially lowered and no remarkable reduction of the apoB mRNA was observed for κ_{min} values >0.22. The correlation between κ_{min} and IC_{50} values on PMHs seems to be more pronounced then on HeLa and RAW264.7 cells.[127] This may contribute to the testing of truly processed and purified liposomes which were dialyzed to remove the organic solvent and non-encapsulated siRNAs, concentrated and analyzed for drug concentration. Liposomes were made by reproducible machine-controlled productions with comparable particle sizes of less than 200 nm. Within this size class, presumably only one liposomal uptake pathway (clathrin-mediated endocytosis) is favored by the cells [163,164] leading to a more definite data plot. κ_{min} is a more necessary determinant for potent transfection events then lipid chemistry (Appendix, Tab. 8.1 & 8.2). Various chemistries of neutral, anionic and cationic lipids were tested differing in tail and head group moieties and molar fractions but their individual κ_{min} value was important and guides the activity of the respective formulations. However, regarding transfection efficiency Amphoter I class liposomes are superior over Amphoter II class liposomes which are composed of charge-reversible cationic lipids characterized by large head group volumes (e.g. imidazole- or morpholine-succineamid moieties) increasing the κ values. Permanently charged cationic lipids of Amphoter I class liposomes contain small head group volumes (e.g. tertiary or quaternary amino moieties) which decrease the κ value and thus benefit the membrane fusion.

The rational design of fusogenic lipid mixtures confirms previous findings of nov038 and gives a potential explanation for the non-fusogenicity of this vector. A FRET-based fusion-assay of nov038 revealed no remarkable pH-induced membrane fusion between these particles (Fig. 3.13 B) and no effective siRNA transfection could be observed on HeLa and RAW264.7 cells.[127] In contrast, formulations with a bi-phasic stability (at pH 4 and pH 7.5) and a substantial fusion propensity (low κ_{min} values) in a weakly acid environment (tested by fusion-assay) indicate an effective cellular transfection, e.g. nov729 (Fig. 3.13). Comparable results were obtained on HeLa and RAW264.7 cells.

Besides siRNAs nov729 also delivers single-stranded ASO molecules (Fig. 3.14). By using ASO molecules transfection efficiency is comparable for both nov729 and nov038. This indicates a vector-assisted ASO delivery by nov038 into the cell which is much more effective than un-assisted delivery of ASO. However, the ASO escape from the endosome is presumably driven by a pH-dependent disintegration of nov038 and subsequent active, protein-guided crossing of the endosomal membrane. Nov729 delivers equally amounts of ASO molecules into the endosome but actively promotes the membrane fusion and cytosolic delivery which is further an imperative for siRNA delivery. This indicates also that the endosome escape which can be directed by selection of fusogenic lipid mixtures (with low κ_{min} values) is more important than the carrier uptake by the cells.

In silico and fusion data support the in vitro findings of nov729 and nov038 in that the latter is not able to fuse with membranes over the entire pH range whereas nov729 is capable to maintain membrane fusion at pH ~5 (Fig. 3.13). Based on fusion data nov038 is not qualified for an effective siRNA delivery and nov729 was chosen for further testing. Selection criteria for nov729 included also the (re-)producibility, stability and reliability during manufacturing and up-scaling process as well as storage stability at a temperature of 4 °C. In contrast to other novel fusogenic candidates (e.g. with somewhat lower IC_{50} values) nov729 formed liposomes with little to no tendency to liposomal aggregation during the production process and storage at physiological pH. Further, nov729 is stable in mouse and human serum (in vitro tests (report nov-001-2009)) and was thus selected for further in vivo testing.

4.8 Delivery of siRNA *in vivo* by using nov729 is inefficient and inhibitable by mouse serum

Consequently, nov729 was used to encapsulate ApoB I 5'P siRNA and prepared with small average particle size to maintain a sufficient hepatic penetration in the mouse ApoB100 model. Application of nov729 in mice was safe and non-toxic (Tab. 3.13) but also did not cause any substantial reduction of apoB100 mRNA levels in the liver (Fig. 3.15). A significant effect could be shown at a dose of 8 mg/kg compared to the scrambled control group. But, total cholesterol and LDL plasma levels were not lowered by the treatment. High lipid and siRNA dosages, however, led to slight, but not significant increase in plasma cholesterol levels compared to the saline group whereas lower doses did not. An additional entry of cholesterol by the liposomal vector (nov729 comprises 40 mol % of cholesterol) may have caused an elevated plasma cholesterol level. An exchange of liposomal cholesterol with plasma lipoproteins is conceivable as well as the uptake of cholesterol-rich liposomes by cells of the MPS [165] or other tissue and endothelia cells. The excess of cellular cholesterol developed from liposomal uptake will be removed by loading of the cholesterol to apolipoproteins, namely ApoA-I on HDL particles.[166]

Discussion

Because nov729 showed an effective cellular siRNA transfection on PMHs but no decisive mRNA knockdown in hepatocytes *in vivo* one could speculate that serum or plasma components affect the integrity of nov729 particles during circulation. Assuming that, once in the tissue, the formulation nov729 is potent enough to transfect cells of the liver parenchyma, the particles were affected on the route from the injection site to the liver. Interactions of liposomes and serum components include the recognition and marking of liposomal particles by serum opsonins for MPS clearance and/or lipid exchange with lipoproteins.[92,111,167,168] *In vitro* studies on HeLa cells indicated an inhibition of the transfection efficiency of nov729 by serum components, most likely lipoproteins, whereas no limiting interactions were visible with proteins of the complement system like C3 or C9 (Appendix, Fig. 8.2).

Therefore, nov729 was tested *in vitro* on PMHs in the presence or absence of 10 % complete mouse serum. The addition of 10 % complete mouse serum had no effect on the apoB100 mRNA expression levels in buffer treated cells. In the absence of mouse serum, as previously shown (Fig. 3.14), nov729d017 mediates a dose-dependent down-regulation of the apoB100 mRNA (Fig. 3.16 A). However, liposomes of nov729 at a siRNA concentration of 100 nM were completely inhibited by the mouse serum and no down-regulation could be determined. Increasing doses of nov729d017 titrated the mouse serum components and led to an mRNA down-regulation (Fig. 3.16 A). Thus, the inhibition depends on the serum component concentration and a high number of nov729 particles are competitive enough to restore the knockdown effect *in vitro*. Conclusively, in mice, in the presence of 100 % serum, a markedly higher amount of nov729 is necessary for an effective liver delivery.

Confocal images support the assumption of an inhibition of nov729 by mouse serum *in vitro*. In the presence of mouse serum the binding and uptake of nov729 was dramatically decreased (Fig. 3.16 B; right image) compared to the non-serum treatment showing a clear staining of the cytoplasm (Fig. 3.16 B; left/central image). Serum components, e.g. lipoproteins, may interfere with nov729 particles hindering them from cellular uptake and thus were washed away from the medium during the washing step. Interference might include a lipid exchange between lipoproteins and liposomes as well as a competition for lipoprotein receptors, e.g. LDL and/or VLDL receptor (LDLR, VLDLR).

Lipid nanoparticles (LNPs) are known to interact with serum proteins, exchanging components and acquiring proteins in circulation that can potentially direct them to specific cell types.[100] Recently, Semple and co-workers mentioned that their lipid nanoparticles (LNPs) bearing an ionizable cationic lipid may associate with proteins in the plasma which promote an enhanced hepatocytes endocytosis after systemic administration.[73] Ionizable lipid nanoparticles (iLNPs) are positively charged at acidic pH but close to charge-neutral at physiological pH. It was found that apolipoproteins adsorb to neutral liposomes but only apolipoprotein E (ApoE) mediates an enhanced uptake into primary hepatocytes [169] whereas ApoE is also involved in the clearance of neutral liposomes by hepatocytes *in vivo*.[170] The acquisition of ApoE by iLNPs during the circulation and delivering of siRNAs to hepatocytes in a targeted manner *in vitro* and *in vivo* was called "endogenous targeting".[171] Akinc and coworkers found that the *in vivo* activity of iLNPs is ApoE dependent whereas ApoE has no effect on the uptake or gene-silencing activity of strictly cationic LNPs (cLNPs).[76] ApoE-targeted iLNPs were internalized by multiple hepatic ApoE receptors and LDLR was found to be an important receptor for the hepatic uptake. ApoE is found on chylomicrons, VLDL and HDL and plays a crucial role in the clearance of VLDL and chylomicron remnants by hepatocytes.[172] Besides LDL receptor mediated endocytosis of particles numerous other receptors including scavenger receptor BI (SR-BI) and LDL related protein (LRP) have been associated with ApoE-mediated uptake by hepatocytes.[173-175] Conclusively, the cellular uptake of iLNPs was considerably diminished in apoE$^{-/-}$ and LDLR$^{-/-}$ knock-out mice and primary hepatocytes compared to the wild-type.[171]

Thus, particles of nov729 are unable to take the cellular uptake pathway of LDLR probably due to a missing interaction with ApoE in the presence of mouse serum. In contrast to iLNPs Smarticles formulation nov729 bears an excess of anionic surface charge which, according to Akinc and co-workers, diminishes the interaction with ApoE.[76] In the presence of 10 % fetal calve serum (FCS) nov729 effectively transfected PMHs which may result from a less competitive environment. Extremely low levels of HDL and LDL and no apparent levels of VLDL in the FCS [176] give an advantage for the liposomal binding to the LDLR (or other receptors) and the subsequent internalization. Mouse serum contains high levels of lipoproteins (HDL>LDL>VLDL) [177,178] and is thus more competitive for binding to lipoprotein receptors on the surface of PMHs or hepatocytes *in vivo*. Seemingly, nov729 particles rely on the cellular uptake caused by lipoprotein receptor mediated endocytosis. A lipid exchange between liposomes and lipoproteins is also imaginable as a modified lipid composition of nov729 may not have the same transfection and fusion properties as the predesigned mixture.

4.9 Conclusions and future perspectives

The effective delivery of oligonucleotides is one of the main challenges in the field of RNA/DNA-based research and therapeutics. Especially in the case of siRNAs the safe and non-toxic transport and delivery to and across the cell membrane is an indispensable requirement for harnessing RNA interference.

Whereas nov038 can be used for an ASO delivery into cells of the MPS [86] and hepatocytes (this work) the formulation was not able to mediate the transport of siRNAs into hepatocytes *in vitro* and *in vivo*. While ASO molecules are thought to be transported across the membrane via receptors or channels, double-stranded and highly polar siRNAs are not and delivery systems have to ensure the secure passage of the intact double helix across the plasma and endosomal membrane.

Novel amphoteric liposomal vectors designed by a rational approach with a pronounced fusion propensity at pH 5-6 mediated the effective crossing and cytosolic delivery of siRNAs in primary mouse hepatocytes. However, the capability of nov729 to deliver siRNAs was totally diminished after systemic administration presumably through an inhibition by mouse lipoproteins and their receptor mediated uptake pathways. Thus, nov729 can be used for efficient siRNA delivery *in vitro*, e.g. primary hepatocytes but has to be improved for *in vivo* delivery in terms of membrane surface properties and cellular uptake.

Tissue-specific surface ligands, for instance, enhance the binding to membrane-bound receptors and uptake by target cells, as recently shown after incorporation of N-acetylgalactosamine (NAG) into LNPs for targeting of the asialoglycoprotein receptor (ASGPR) on hepatocytes.[171] The receptor-mediated endocytosis of NAG-coupled LNPs improved the RNAi-mediated knockdown of a hepatic mRNA and protein by more than 10 times. However, the ligandation of nanoparticles complicates the delivery system, increases manufacturing costs and may have a negative lasting impact on the immune system.

Here, the targeted design of lipid head groups with selective binding properties displays a chemical and biological alternative. Lipid head groups and lipid compositions can be designed for an improved interaction with plasma lipoproteins (e.g. ApoE) or the direct and competitive association with membrane-bound lipoprotein receptors (e.g. LDLR, SR-BI). In the case of receptor interactions cationic surface charges which are in connection to the chemical structure of the cationic amino acids arginine or lysine would benefit, for instance, the affinity to the LDLR. A distinct sequence of arginine and lysine within the ApoE lipoprotein is responsible for the selective binding to the LDLR and the subsequent receptor-mediated endocytosis of the complex.

Finally, the development of lipid nanoparticles which avoid uptake by the "classical" endocytotic pathway including clathrin or caveolin-mediated endocytosis is of special interest for the drug delivery field. Love and co-workers recently proposed a non-LDLR-mediated uptake of their LNPs containing an epoxide-derived lipid-like (lipidoid) compound (C12-200) by macropinocytosis.[78] Macropinosomes are thought not to merge with the degradative pathway and LNPs thus avoid the lysosomal degradation often encountered with endocytosis.[164] LNPs comprising C12-200 showed a several hundredfold improvement in potency in mice compared to LNPs of the previous generation. For therapeutic perspectives in humans, lowering the injected siRNA dose also comes along with a concomitant reduction in dosed formulation excipients.[78]

5. Summary

Amphoteric liposomes contain lipids bearing pH-sensitive elements which sense the environmental pH. Under acidic conditions they become cationic and thus stably sequester nucleic acids. At physiological pH amphoteric liposomes are negatively charged and thus prevent the aggregation with anionic serum components during circulation. Optimized amphoteric liposomes provide a rational mechanism for the pH dependent fusion of the liposomal and endosomal membrane. They can therefore be used for the cytosolic delivery of oligonucleotides. Liposomes specifically address challenges involved with the transit of oligonucleotides into cells, namely biodistribution, cellular uptake and endosomal release, and are expected to unleash the full potential of oligonucleotide-based therapeutics.

Firstly, this thesis placed emphasis on the pharmacokinetic (PK) and biodistribution (BD) of the amphoteric formulation nov038 employing a Cy5.5-labeled antisense molecule (ASO) for tracking purposes. Nov038 liposomes show a non-linear, dose-dependent PK in which $t_{1/2}$, C_{max} and Cl_{tot} alter in a dose-related, but not dose-linear fashion. Increasing lipid doses of nov038 lead to a saturation of the first compartment (mainly cells of the MPS) and second compartment (e.g. liver parenchyma) facilitating a prolonged blood circulation. Low doses of nov038 distribute almost completely into liver and spleen whereas higher lipid doses presumably enable the distribution to peripheral body sites. Uptake of the formulated Cy5.5-labeled ASO by hepatocytes was confirmed by CLSM and in a following pharmacodynamic study nov038 mediated the knockdown of a hepatic mRNA (LT1) using formulated LT1-ASO in a dose dependent manner. Moreover, nov038 potentiated the antisense effect compared to free, non-encapsulated, ASO. However, nov038 was not able to deliver double-stranded siRNA molecules since a reduction of the apoB100 mRNA or protein by using ApoB I siRNA could not be detected, neither *in vivo* nor *in vitro*.

The hypothesis was thus that a fusion of liposomes with the endosomal membrane is a prerequisite for the intracellular release of siRNAs, whereas ASO molecules have been shown to cross membranes independent of supporting agents. The second part of this thesis aimed at the development of novel fusogenic liposomal formulations created by a rational design approach. The parameter describing the lipid shape in particular consideration of the environmental pH and counterion binding, κ_{min}, was introduced as a valuable prediction factor for fusogenicity. In fact, formulations with a small κ_{min} value (e.g. nov729) were able to mediate a siRNA transfection with low IC_{50} values on primary mouse hepatocytes. Unfortunately, the siRNA-mediated reduction of a hepatic mRNA *in vivo* could not be shown. Most likely, particles of nov729 share the LDL receptor uptake pathway and thus compete with lipoproteins. Further development is necessary to overcome this inhibition by lipoproteins and to mediate the delivery of siRNA molecules *in vivo*.

6. References

1. **Cerritelli, S. M. and R. J. Crouch.** 2009. Ribonuclease H: the enzymes in eukaryotes. FEBS J. **276**:1494-1505.

2. **Bennett, C. F. and E. E. Swayze.** 2010. RNA targeting therapeutics: molecular mechanisms of antisense oligonucleotides as a therapeutic platform. Annu. Rev. Pharmacol. Toxicol. **50**:259-293.

3. **Kurreck, J.** 2003. Antisense technologies. Improvement through novel chemical modifications. Eur. J. Biochem. **270**:1628-1644.

4. **Herdewijn, P.** 2000. Heterocyclic modifications of oligonucleotides and antisense technology. Antisense Nucleic Acid Drug Dev. **10**:297-310.

5. **Campbell, J. M., T. A. Bacon, and E. Wickstrom.** 1990. Oligodeoxynucleoside phosphorothioate stability in subcellular extracts, culture media, sera and cerebrospinal fluid. J. Biochem. Biophys. Methods **20**:259-267.

6. **Phillips, M. I. and Y. C. Zhang.** 2000. Basic principles of using antisense oligonucleotides in vivo. Methods Enzymol. **313**:46-56.

7. **Brown, D. A., S. H. Kang, S. M. Gryaznov, L. DeDionisio, O. Heidenreich, S. Sullivan, X. Xu, and M. I. Nerenberg.** 1994. Effect of phosphorothioate modification of oligodeoxynucleotides on specific protein binding. J. Biol. Chem. **269**:26801-26805.

8. **Levin, A. A.** 1999. A review of the issues in the pharmacokinetics and toxicology of phosphorothioate antisense oligonucleotides. Biochim. Biophys. Acta **1489**:69-84.

9. **Geary, R. S., R. Z. Yu, and A. A. Levin.** 2001. Pharmacokinetics of phosphorothioate antisense oligodeoxynucleotides. Curr. Opin. Investig. Drugs **2**:562-573.

10. **Furdon, P. J., Z. Dominski, and R. Kole.** 1989. RNase H cleavage of RNA hybridized to oligonucleotides containing methylphosphonate, phosphorothioate and phosphodiester bonds. Nucleic Acids Res. **17**:9193-9204.

11. **Henry, S. P., P. C. Giclas, J. Leeds, M. Pangburn, C. Auletta, A. A. Levin, and D. J. Kornbrust.** 1997. Activation of the alternative pathway of complement by a phosphorothioate oligonucleotide: potential mechanism of action. J. Pharmacol. Exp. Ther. **281**:810-816.

12. **Agrawal, S. and Q. Zhao.** 1998. Antisense therapeutics. Curr. Opin. Chem. Biol. **2**:519-528.

13. **Jason, T. L., J. Koropatnick, and R. W. Berg.** 2004. Toxicology of antisense therapeutics. Toxicol. Appl. Pharmacol. **201**:66-83.

14. **Temsamani, J., A. Roskey, C. Chaix, and S. Agrawal.** 1997. In vivo metabolic profile of a phosphorothioate oligodeoxyribonucleotide. Antisense Nucleic Acid Drug Dev. **7**:159-165.

15. **Crooke, R. M., M. J. Graham, M. J. Martin, K. M. Lemonidis, T. Wyrzykiewicz, and L. L. Cummins.** 2000. Metabolism of antisense oligonucleotides in rat liver homogenates. J. Pharmacol. Exp. Ther. **292**:140-149.

16. **Agrawal, S., J. Temsamani, and J. Y. Tang.** 1991. Pharmacokinetics, biodistribution, and stability of oligodeoxynucleotide phosphorothioates in mice. Proc. Natl. Acad. Sci. U. S. A **88**:7595-7599.

17. **Cummins, L. L., S. R. Owens, L. M. Risen, E. A. Lesnik, S. M. Freier, D. McGee, C. J. Guinosso, and P. D. Cook.** 1995. Characterization of fully 2'-modified oligoribonucleotide hetero- and homoduplex hybridization and nuclease sensitivity. Nucleic Acids Res. **23**:2019-2024.

18. **Freier, S. M. and K. H. Altmann.** 1997. The ups and downs of nucleic acid duplex stability: structure-stability studies on chemically-modified DNA:RNA duplexes. Nucleic Acids Res. **25**:4429-4443.

19. **Geary, R. S., O. Khatsenko, K. Bunker, R. Crooke, M. Moore, T. Burckin, L. Truong, H. Sasmor, and A. A. Levin.** 2001. Absolute bioavailability of 2'-O-(2-methoxyethyl)-modified antisense oligonucleotides following intraduodenal instillation in rats. J. Pharmacol. Exp. Ther. **296**:898-904.

References

20. **Geary, R. S.** 2009. Antisense oligonucleotide pharmacokinetics and metabolism. Expert. Opin. Drug Metab Toxicol. **5**:381-391.

21. **Yu, R. Z., R. S. Geary, D. K. Monteith, J. Matson, L. Truong, J. Fitchett, and A. A. Levin.** 2004. Tissue disposition of 2'-O-(2-methoxy) ethyl modified antisense oligonucleotides in monkeys. J. Pharm. Sci. **93**:48-59.

22. **Kurreck, J., E. Wyszko, C. Gillen, and V. A. Erdmann.** 2002. Design of antisense oligonucleotides stabilized by locked nucleic acids. Nucleic Acids Res. **30**:1911-1918.

23. **Fluiter, K., M. Frieden, J. Vreijling, C. Rosenbohm, M. B. de Wissel, S. M. Christensen, T. Koch, H. Orum, and F. Baas.** 2005. On the in vitro and in vivo properties of four locked nucleic acid nucleotides incorporated into an anti-H-Ras antisense oligonucleotide. Chembiochem. **6**:1104-1109.

24. **Geary, R. S., J. Matson, and A. A. Levin.** 1999. A nonradioisotope biomedical assay for intact oligonucleotide and its chain-shortened metabolites used for determination of exposure and elimination half-life of antisense drugs in tissue. Anal. Biochem. **274**:241-248.

25. **Zhang, H., J. Cook, J. Nickel, R. Yu, K. Stecker, K. Myers, and N. M. Dean.** 2000. Reduction of liver Fas expression by an antisense oligonucleotide protects mice from fulminant hepatitis. Nat. Biotechnol. **18**:862-867.

26. **Monia, B. P., E. A. Lesnik, C. Gonzalez, W. F. Lima, D. McGee, C. J. Guinosso, A. M. Kawasaki, P. D. Cook, and S. M. Freier.** 1993. Evaluation of 2'-modified oligonucleotides containing 2'-deoxy gaps as antisense inhibitors of gene expression. J. Biol. Chem. **268**:14514-14522.

27. **Pan, W. H. and G. A. Clawson.** 2006. Antisense applications for biological control. J. Cell Biochem. **98**:14-35.

28. **Braasch, D. A. and D. R. Corey.** 2001. Locked nucleic acid (LNA): fine-tuning the recognition of DNA and RNA. Chem. Biol. **8**:1-7.

29. **Bondensgaard, K., M. Petersen, S. K. Singh, V. K. Rajwanshi, R. Kumar, J. Wengel, and J. P. Jacobsen.** 2000. Structural studies of LNA:RNA duplexes by NMR: conformations and implications for RNase H activity. Chemistry. **6**:2687-2695.

30. **Fire, A., S. Xu, M. K. Montgomery, S. A. Kostas, S. E. Driver, and C. C. Mello.** 1998. Potent and specific genetic interference by double-stranded RNA in Caenorhabditis elegans. Nature **391**:806-811.

31. **Hannon, G. J. and J. J. Rossi.** 2004. Unlocking the potential of the human genome with RNA interference. Nature **431**:371-378.

32. **Elbashir, S. M., W. Lendeckel, and T. Tuschl.** 2001. RNA interference is mediated by 21- and 22- nucleotide RNAs. Genes Dev. **15**:188-200.

33. **Bernstein, E., A. M. Denli, and G. J. Hannon.** 2001. The rest is silence. RNA. **7**:1509-1521.

34. **Zamore, P. D., T. Tuschl, P. A. Sharp, and D. P. Bartel.** 2000. RNAi: double-stranded RNA directs the ATP-dependent cleavage of mRNA at 21 to 23 nucleotide intervals. Cell **101**:25-33.

35. **Hammond, S. M., E. Bernstein, D. Beach, and G. J. Hannon.** 2000. An RNA-directed nuclease mediates post-transcriptional gene silencing in Drosophila cells. Nature **404**:293-296.

36. **Nykanen, A., B. Haley, and P. D. Zamore.** 2001. ATP requirements and small interfering RNA structure in the RNA interference pathway. Cell **107**:309-321.

37. **Wang, Y., S. Juranek, H. Li, G. Sheng, T. Tuschl, and D. J. Patel.** 2008. Structure of an argonaute silencing complex with a seed-containing guide DNA and target RNA duplex. Nature **456**:921-926.

38. **Aagaard, L. and J. J. Rossi.** 2007. RNAi therapeutics: principles, prospects and challenges. Adv. Drug Deliv. Rev. **59**:75-86.

39. **Bertrand, J. R., M. Pottier, A. Vekris, P. Opolon, A. Maksimenko, and C. Malvy.** 2002. Comparison of antisense oligonucleotides and siRNAs in cell culture and in vivo. Biochem. Biophys. Res. Commun. **296**:1000-1004.

References

40. **Dykxhoorn, D. M. and J. Lieberman**. 2005. The Silent Revolution: RNA Interference as Basic Biology, Research Tool, and Therapeutic. Annual Review of Medicine **56**:401-423.

41. **Haupenthal, J., C. Baehr, S. Zeuzem, and A. Piiper**. 2007. RNAse A-like enzymes in serum inhibit the anti-neoplastic activity of siRNA targeting polo-like kinase 1. Int. J. Cancer **121**:206-210.

42. **Braasch, D. A., Z. Paroo, A. Constantinescu, G. Ren, O. K. Oz, R. P. Mason, and D. R. Corey**. 2004. Biodistribution of phosphodiester and phosphorothioate siRNA. Bioorg. Med. Chem. Lett. **14**:1139-1143.

43. **Czauderna, F., M. Fechtner, S. Dames, H. Aygun, A. Klippel, G. J. Pronk, K. Giese, and J. Kaufmann**. 2003. Structural variations and stabilising modifications of synthetic siRNAs in mammalian cells. Nucleic Acids Res. **31**:2705-2716.

44. **Mook, O. R., F. Baas, M. B. de Wissel, and K. Fluiter**. 2007. Evaluation of locked nucleic acid-modified small interfering RNA in vitro and in vivo. Mol. Cancer Ther. **6**:833-843.

45. **Jackson, A. L., J. Burchard, J. Schelter, B. N. Chau, M. Cleary, L. Lim, and P. S. Linsley**. 2006. Widespread siRNA 'off-target' transcript silencing mediated by seed region sequence complementarity. RNA **12**:1179-1187.

46. **Khvorova, A., A. Reynolds, and S. D. Jayasena**. 2003. Functional siRNAs and miRNAs Exhibit Strand Bias. Cell **115**:209-216.

47. **Jackson, A. L., J. Burchard, D. Leake, A. Reynolds, J. Schelter, J. Guo, J. M. Johnson, L. Lim, J. Karpilow, K. Nichols, W. Marshall, A. Khvorova, and P. S. Linsley**. 2006. Position-specific chemical modification of siRNAs reduces "off-target" transcript silencing. RNA. **12**:1197-1205.

48. **Sioud, M**. 2005. Induction of inflammatory cytokines and interferon responses by double-stranded and single-stranded siRNAs is sequence-dependent and requires endosomal localization. J. Mol. Biol. **348**:1079-1090.

49. **Gantier, M. P. and B. R. G. Williams**. 2007. The response of mammalian cells to double-stranded RNA. Cytokine & Growth Factor Reviews **18**:363-371.

50. **Marques, J. T. and B. R. Williams**. 2005. Activation of the mammalian immune system by siRNAs. Nat. Biotechnol. **23**:1399-1405.

51. **Judge, A. D., V. Sood, J. R. Shaw, D. Fang, K. McClintock, and I. MacLachlan**. 2005. Sequence-dependent stimulation of the mammalian innate immune response by synthetic siRNA. Nat Biotech **23**:457-462.

52. **Takeda, K. and S. Akira**. 2005. Toll-like receptors in innate immunity. Int. Immunol. **17**:1-14.

53. **Heil, F., H. Hemmi, H. Hochrein, F. Ampenberger, C. Kirschning, S. Akira, G. Lipford, H. Wagner, and S. Bauer**. 2004. Species-specific recognition of single-stranded RNA via toll-like receptor 7 and 8. Science **303**:1526-1529.

54. **Judge, A. D., G. Bola, A. C. Lee, and I. MacLachlan**. 2006. Design of noninflammatory synthetic siRNA mediating potent gene silencing in vivo. Mol. Ther. **13**:494-505.

55. **Yakubov, L. A., E. A. Deeva, V. F. Zarytova, E. M. Ivanova, A. S. Ryte, L. V. Yurchenko, and V. V. Vlassov**. 1989. Mechanism of oligonucleotide uptake by cells: involvement of specific receptors? Proc. Natl. Acad. Sci. U. S. A **86**:6454-6458.

56. **Akhtar, S., S. Basu, E. Wickstrom, and R. L. Juliano**. 1991. Interactions of antisense DNA oligonucleotide analogs with phospholipid membranes (liposomes). Nucleic Acids Res. **19**:5551-5559.

57. **Stein, C. A., J. B. Hansen, J. Lai, S. Wu, A. Voskresenskiy, A. Hog, J. Worm, M. Hedtjarn, N. Souleimanian, P. Miller, H. S. Soifer, D. Castanotto, L. Benimetskaya, H. Orum, and T. Koch**. 2010. Efficient gene silencing by delivery of locked nucleic acid antisense oligonucleotides, unassisted by transfection reagents. Nucleic Acids Res. **38**:e3.

58. **Gupta, N., N. Fisker, M. C. Asselin, M. Lindholm, C. Rosenbohm, H. Orum, J. Elmen, N. G. Seidah, and E. M. Straarup**. 2010. A Locked Nucleic Acid Antisense Oligonucleotide (LNA) Silences PCSK9 and Enhances LDLR Expression in vitro and in vivo. PLoS ONE **5**:e10682.

References

59. **Kastelein, J. J. P., M. K. Wedel, B. F. Baker, J. Su, J. D. Bradley, R. Z. Yu, E. Chuang, M. J. Graham, and R. M. Crooke.** 2006. Potent Reduction of Apolipoprotein B and Low-Density Lipoprotein Cholesterol by Short-Term Administration of an Antisense Inhibitor of Apolipoprotein B. Circulation **114**:1729-1735.

60. **Kling, J.** 2010. Safety signal dampens reception for mipomersen antisense. Nat Biotech **28**:295-297.

61. **Lewis, D. L. and J. A. Wolff.** 2007. Systemic siRNA delivery via hydrodynamic intravascular injection. Adv. Drug Deliv. Rev. **59**:115-123.

62. **McCaffrey, A. P., L. Meuse, T. T. Pham, D. S. Conklin, G. J. Hannon, and M. A. Kay.** 2002. RNA interference in adult mice. Nature **418**:38-39.

63. **Feinberg, E. H. and C. P. Hunter.** 2003. Transport of dsRNA into cells by the transmembrane protein SID-1. Science **301**:1545-1547.

64. **Winston, W. M., C. Molodowitch, and C. P. Hunter.** 2002. Systemic RNAi in C. elegans Requires the Putative Transmembrane Protein SID-1. Science **295**:2456-2459.

65. **Shih, J. D., M. C. Fitzgerald, M. Sutherlin, and C. P. Hunter.** 2009. The SID-1 double-stranded RNA transporter is not selective for dsRNA length. RNA **15**:384-390.

66. **Duxbury, M. S., S. W. Ashley, and E. E. Whang.** 2005. RNA interference: A mammalian SID-1 homologue enhances siRNA uptake and gene silencing efficacy in human cells. Biochemical and Biophysical Research Communications **331**:459-463.

67. **Wolfrum, C., S. Shi, K. N. Jayaprakash, M. Jayaraman, G. Wang, R. K. Pandey, K. G. Rajeev, T. Nakayama, K. Charrise, E. M. Ndungo, T. Zimmermann, V. Koteliansky, M. Manoharan, and M. Stoffel.** 2007. Mechanisms and optimization of in vivo delivery of lipophilic siRNAs. Nat. Biotechnol. **25**:1149-1157.

68. **Watts, J. K. and D. R. Corey.** 2010. Clinical status of duplex RNA. Bioorg. Med. Chem. Lett. **20**:3203-3207.

69. **Schreiber, S., S. Nikolaus, H. Malchow, W. Kruis, H. Lochs, A. Raedler, E. G. Hahn, T. Krummenerl, and G. Steinmann.** 2001. Absence of efficacy of subcutaneous antisense ICAM-1 treatment of chronic active Crohn's disease. Gastroenterology **120**:1339-1346.

70. **Reinsch, C., E. Siepi, A. Dieckmann, and S. Panzner.** 2008. Strategies for the Delivery of Oligonucleotides in vivo, p. 226-240. In J. Kurreck (ed.), Therapeutic Oligonucleotides. The Royal Society of Chemistry, Cambridge, UK.

71. **Reischl, D. and A. Zimmer.** 2009. Drug delivery of siRNA therapeutics: potentials and limits of nanosystems. Nanomedicine. **5**:8-20.

72. **Wu, S. Y. and N. A. McMillan.** 2009. Lipidic systems for in vivo siRNA delivery. AAPS. J. **11**:639-652.

73. **Semple, S. C., A. Akinc, J. Chen, A. P. Sandhu, B. L. Mui, C. K. Cho, D. W. Y. Sah, D. Stebbing, E. J. Crosley, E. Yaworski, I. M. Hafez, J. R. Dorkin, J. Qin, K. Lam, K. G. Rajeev, K. F. Wong, L. B. Jeffs, L. Nechev, M. L. Eisenhardt, M. Jayaraman, M. Kazem, M. A. Maier, M. Srinivasulu, M. J. Weinstein, Q. Chen, R. Alvarez, S. A. Barros, S. De, S. K. Klimuk, T. Borland, V. Kosovrasti, W. L. Cantley, Y. K. Tam, M. Manoharan, M. A. Ciufolini, M. A. Tracy, A. de Fougerolles, I. MacLachlan, P. R. Cullis, T. D. Madden, and M. J. Hope.** 2010. Rational design of cationic lipids for siRNA delivery. Nat Biotech **28**:172-176.

74. **Zhang, Y. P., L. Sekirov, E. G. Saravolac, J. J. Wheeler, P. Tardi, K. Clow, E. Leng, R. Sun, P. R. Cullis, and P. Scherrer.** 1999. Stabilized plasmid-lipid particles for regional gene therapy: formulation and transfection properties. Gene Ther. **6**:1438-1447.

75. **Akinc, A., A. Zumbuehl, M. Goldberg, E. S. Leshchiner, V. Busini, N. Hossain, S. A. Bacallado, D. N. Nguyen, J. Fuller, R. Alvarez, A. Borodovsky, T. Borland, R. Constien, F. A. de, J. R. Dorkin, J. K. Narayanannair, M. Jayaraman, M. John, V. Koteliansky, M. Manoharan, L. Nechev, J. Qin, T. Racie, D. Raitcheva, K. G. Rajeev, D. W. Sah, J. Soutschek, I. Toudjarska, H. P. Vornlocher, T. S. Zimmermann, R. Langer, and D. G. Anderson.** 2008. A combinatorial library of lipid-like materials for delivery of RNAi therapeutics. Nat. Biotechnol.

References

76. Akinc, A., M. Goldberg, J. Qin, J. R. Dorkin, C. Gamba-Vitalo, M. Maier, K. N. Jayaprakash, M. Jayaraman, K. G. Rajeev, M. Manoharan, V. Koteliansky, I. Rohl, E. S. Leshchiner, R. Langer, and D. G. Anderson. 2009. Development of Lipidoid-siRNA Formulations for Systemic Delivery to the Liver. Mol Ther **17**:872-879.

77. Frank-Kamenetsky, M., A. Grefhorst, N. N. Anderson, T. S. Racie, B. Bramlage, A. Akinc, D. Butler, K. Charisse, R. Dorkin, Y. Fan, C. Gamba-Vitalo, P. Hadwiger, M. Jayaraman, M. John, K. N. Jayaprakash, M. Maier, L. Nechev, K. G. Rajeev, T. Read, I. Rohl, J. Soutschek, P. Tan, J. Wong, G. Wang, T. Zimmermann, A. de Fougerolles, H. P. Vornlocher, R. Langer, D. G. Anderson, M. Manoharan, V. Koteliansky, J. D. Horton, and K. Fitzgerald. 2008. Therapeutic RNAi targeting PCSK9 acutely lowers plasma cholesterol in rodents and LDL cholesterol in nonhuman primates. Proceedings of the National Academy of Sciences **105**:11915-11920.

78. Love, K. T., K. P. Mahon, C. G. Levins, K. A. Whitehead, W. Querbes, J. R. Dorkin, J. Qin, W. Cantley, L. L. Qin, T. Racie, M. Frank-Kamenetsky, K. N. Yip, R. Alvarez, D. W. Sah, F. A. de, K. Fitzgerald, V. Koteliansky, A. Akinc, R. Langer, and D. G. Anderson. 2010. Lipid-like materials for low-dose, in vivo gene silencing. Proc. Natl. Acad. Sci. U. S. A .

79. Halder, J., A. A. Kamat, C. N. Landen, Jr., L. Y. Han, S. K. Lutgendorf, Y. G. Lin, W. M. Merritt, N. B. Jennings, A. Chavez-Reyes, R. L. Coleman, D. M. Gershenson, R. Schmandt, S. W. Cole, G. Lopez-Berestein, and A. K. Sood. 2006. Focal adhesion kinase targeting using in vivo short interfering RNA delivery in neutral liposomes for ovarian carcinoma therapy. Clin. Cancer Res. **12**:4916-4924.

80. Landen, C. N., Jr., A. Chavez-Reyes, C. Bucana, R. Schmandt, M. T. Deavers, G. Lopez-Berestein, and A. K. Sood. 2005. Therapeutic EphA2 gene targeting in vivo using neutral liposomal small interfering RNA delivery. Cancer Res. **65**:6910-6918.

81. Klimuk, S. K., S. C. Semple, P. N. Nahirney, M. C. Mullen, C. F. Bennett, P. Scherrer, and M. J. Hope. 2000. Enhanced anti-inflammatory activity of a liposomal intercellular adhesion molecule-1 antisense oligodeoxynucleotide in an acute model of contact hypersensitivity. J. Pharmacol. Exp. Ther. **292**:480-488.

82. Collins, D. 1995. pH-sensitive liposomes as tools for cytoplasmic delivery, p. 201-214. In J. R. Philippot and F. Schubert (ed.), Liposomes as tools in basic research and industry. CRC Press, Boca Raton.

83. Hafez, I. M., S. Ansell, and P. R. Cullis. 2000. Tunable pH-sensitive liposomes composed of mixtures of cationic and anionic lipids. Biophys. J. **79**:1438-1446.

84. Panzner, S., S. Fankhänel, F. Essler, and C. Panzner. 2002. Amphoteric liposomes and the use thereof. Halle/S., DE patent WO 02/066012.

85. Gao, D., A. H. Wagner, S. Fankhaenel, T. Stojanovic, S. Schweyer, S. Panzner, and M. Hecker. 2005. CD40 antisense oligonucleotide inhibition of trinitrobenzene sulphonic acid induced rat colitis. Gut **54**:70-77.

86. Andreakos, E., U. Rauchhaus, A. Stavropoulos, G. Endert, V. Wendisch, A. S. Benahmed, S. Giaglis, J. Karras, S. Lee, H. Gaus, C. F. Bennett, R. O. Williams, P. Sideras, and S. Panzner. 2009. Amphoteric liposomes enable systemic antigen-presenting cell-directed delivery of CD40 antisense and are therapeutically effective in experimental arthritis. Arthritis Rheum. **60**:994-1005.

87. Reinsch, C., U. Rauchhaus, and S. Panzner. 2008. Amphoteric liposomes are a platform for multi-organ delivery of oligonucleotides, p. 328-331. In Technical Proceedings of the 2008 NSTI Nanotechnology Conference and Trade Show.

88. Abra, R. M. and C. A. Hunt. 1981. Liposome disposition in vivo. III. Dose and vesicle-size effects. Biochim. Biophys. Acta **666**:493-503.

89. Rahman, Y. E., E. A. Cerny, K. R. Patel, E. H. Lau, and B. J. Wright. 1982. Differential uptake of liposomes varying in size and lipid composition by parenchymal and kupffer cells of mouse liver. Life Sci. **31**:2061-2071.

90. Senior, J. H. 1987. Fate and behavior of liposomes in vivo: a review of controlling factors. Crit Rev. Ther. Drug Carrier Syst. **3**:123-193.

91. Allen, T. M. and C. Hansen. 1991. Pharmacokinetics of stealth versus conventional liposomes: effect of dose. Biochim. Biophys. Acta **1068**:133-141.

References

92. **Hernandez-Caselles, T., J. Villalain, and J. C. Gomez-Fernandez**. 1993. Influence of liposome charge and composition on their interaction with human blood serum proteins. Mol. Cell Biochem. **120**:119-126.

93. **Litzinger, D. C., A. M. Buiting, R. N. van, and L. Huang**. 1994. Effect of liposome size on the circulation time and intraorgan distribution of amphipathic poly(ethylene glycol)-containing liposomes. Biochim. Biophys. Acta **1190**:99-107.

94. **Allen, T. M.** 1988. Toxicity of drug carriers to the mononuclear phagocyte system. Advanced Drug Delivery Reviews **2**:55-67.

95. **Thurston, G., J. W. McLean, M. Rizen, P. Baluk, A. Haskell, T. J. Murphy, D. Hanahan, and D. M. McDonald**. 1998. Cationic liposomes target angiogenic endothelial cells in tumors and chronic inflammation in mice. J. Clin. Invest **101**:1401-1413.

96. **Krasnici, S., A. Werner, M. E. Eichhorn, M. Schmitt-Sody, S. A. Pahernik, B. Sauer, B. Schulze, M. Teifel, U. Michaelis, K. Naujoks, and M. Dellian**. 2003. Effect of the surface charge of liposomes on their uptake by angiogenic tumor vessels. Int. J. Cancer **105**:561-567.

97. **Mahato, R. I., K. Anwer, F. Tagliaferri, C. Meaney, P. Leonard, M. S. Wadhwa, M. Logan, M. French, and A. Rolland**. 1998. Biodistribution and Gene Expression of Lipid/Plasmid Complexes after Systemic Administration. Human Gene Therapy **9**:2083-2099.

98. **Garbuzenko, O., M. Saad, S. Betigeri, M. Zhang, A. Vetcher, V. Soldatenkov, D. Reimer, V. Pozharov, and T. Minko**. 2009. Intratracheal Versus Intravenous Liposomal Delivery of siRNA, Antisense Oligonucleotides and Anticancer Drug. Pharmaceutical Research **26**:382-394.

99. **Dellian, M., F. Yuan, V. S. Trubetskoy, V. P. Torchilin, and R. K. Jain**. 2000. Vascular permeability in a human tumour xenograft: molecular charge dependence. Br. J. Cancer **82**:1513-1518.

100. **Cullis, P. R., A. Chonn, and S. C. Semple**. 1998. Interactions of liposomes and lipid-based carrier systems with blood proteins: Relation to clearance behaviour in vivo. Adv. Drug Deliv. Rev. **32**:3-17.

101. **Szebeni, J.** 1998. The interaction of liposomes with the complement system. Crit Rev. Ther. Drug Carrier Syst. **15**:57-88.

102. **Woodle, M. C. and D. D. Lasic**. 1992. Sterically stabilized liposomes. Biochim. Biophys. Acta **1113**:171-199.

103. **Meyer, O., D. Kirpotin, K. Hong, B. Sternberg, J. W. Park, M. C. Woodle, and D. Papahadjopoulos**. 1998. Cationic Liposomes Coated with Polyethylene Glycol As Carriers for Oligonucleotides. J. Biol. Chem. **273**:15621-15627.

104. **Nicolazzi, C., N. Mignet, F. N. de la, M. Cadet, R. T. Ibad, J. Seguin, D. Scherman, and M. Bessodes**. 2003. Anionic polyethyleneglycol lipids added to cationic lipoplexes increase their plasmatic circulation time. J. Control Release **88**:429-443.

105. **Awasthi, V. D., D. Garcia, R. Klipper, B. A. Goins, and W. T. Phillips**. 2004. Neutral and anionic liposome-encapsulated hemoglobin: effect of postinserted poly(ethylene glycol)-distearoylphosphatidylethanolamine on distribution and circulation kinetics. J. Pharmacol. Exp. Ther. **309**:241-248.

106. **Dams, E. T., P. Laverman, W. J. Oyen, G. Storm, G. L. Scherphof, J. W. van Der Meer, F. H. Corstens, and O. C. Boerman**. 2000. Accelerated blood clearance and altered biodistribution of repeated injections of sterically stabilized liposomes. J. Pharmacol. Exp. Ther. **292**:1071-1079.

107. **Semple, S. C., T. O. Harasym, K. A. Clow, S. M. Ansell, S. K. Klimuk, and M. J. Hope**. 2005. Immunogenicity and rapid blood clearance of liposomes containing polyethylene glycol-lipid conjugates and nucleic Acid. J. Pharmacol. Exp. Ther. **312**:1020-1026.

108. **Yuan, F., M. Dellian, D. Fukumura, M. Leunig, D. A. Berk, V. P. Torchilin, and R. K. Jain**. 1995. Vascular permeability in a human tumor xenograft: molecular size dependence and cutoff size. Cancer Res. **55**:3752-3756.

109. **Harashima, H., T. M. Huong, T. Ishida, Y. Manabe, H. Matsuo, and H. Kiwada**. 1996. Synergistic effect between size and cholesterol content in the enhanced hepatic uptake clearance of liposomes through complement activation in rats. Pharm. Res. **13**:1704-1709.

110. **Devine, D. V., K. Wong, K. Serrano, A. Chonn, and P. R. Cullis**. 1994. Liposome-complement interactions in rat serum: implications for liposome survival studies. Biochim. Biophys. Acta **1191**:43-51.

111. **Ishida, T., H. Harashima, and H. Kiwada**. 2002. Liposome clearance. Biosci. Rep. **22**:197-224.

112. **Scherphof, G. L. and J. A. Kamps**. 2001. The role of hepatocytes in the clearance of liposomes from the blood circulation. Prog. Lipid Res. **40**:149-166.

113. **Roerdink, F., J. Dijkstra, G. Hartman, B. Bolscher, and G. Scherphof**. 1981. The involvement of parenchymal, Kupffer and endothelial liver cells in the hepatic uptake of intravenously injected liposomes. Effects of lanthanum and gadolinium salts. Biochim. Biophys. Acta **677**:79-89.

114. **Wisse, E.** 1970. An electron microscopic study of the fenestrated endothelial lining of rat liver sinusoids. J. Ultrastruct. Res. **31**:125-150.

115. **Oku, N. and Y. Namba**. 1994. Long-circulating liposomes. Crit Rev. Ther. Drug Carrier Syst. **11**:231-270.

116. **Davis, M. E., J. E. Zuckerman, C. H. Choi, D. Seligson, A. Tolcher, C. A. Alabi, Y. Yen, J. D. Heidel, and A. Ribas**. 2010. Evidence of RNAi in humans from systemically administered siRNA via targeted nanoparticles. Nature **464**:1067-1070.

117. **Mayer, L. D., M. B. Bally, P. R. Cullis, S. L. Wilson, and J. T. Emerman**. 1990. Comparison of free and liposome encapsulated doxorubicin tumor drug uptake and antitumor efficacy in the SC115 murine mammary tumor. Cancer Lett. **53**:183-190.

118. **Oja, C. D., S. C. Semple, A. Chonn, and P. R. Cullis**. 1996. Influence of dose on liposome clearance: critical role of blood proteins. Biochim. Biophys. Acta **1281**:31-37.

119. **Chow, D. D., H. E. Essien, M. M. Padki, and K. J. Hwang**. 1989. Targeting small unilamellar liposomes to hepatic parenchymal cells by dose effect. J. Pharmacol. Exp. Ther. **248**:506-513.

120. **Israelachvili, J. N., S. Marcelja, and R. G. Horn**. 1980. Physical principles of membrane organization. Q. Rev. Biophys. **13**:121-200.

121. **Cullis, P. R., D. Fenske, and M. J. Hope**. 1996. Physical properties and functional roles of lipids in membranes, p. 1-33. In D. E. Vance and J. E. Vance (ed.), Biochemistry of Lipids, Lipoproteins and Membranes. Elsevier Science B.V.

122. **Mouritsen, O. G.** 2005. Life - as a matter of fat. Springer Verlag, Berlin Heidelberg.

123. **Fattal, E., P. Couvreur, and C. Dubernet**. 2004. "Smart" delivery of antisense oligonucleotides by anionic pH-sensitive liposomes. Adv. Drug Deliv. Rev. **56**:931-946.

124. **Simoes, S., J. N. Moreira, C. Fonseca, N. Duzgunes, and M. C. de Lima**. 2004. On the formulation of pH-sensitive liposomes with long circulation times. Adv. Drug Deliv. Rev. **56**:947-965.

125. **Li, X. and M. Schick**. 2001. Theory of tunable pH-sensitive vesicles of anionic and cationic lipids or anionic and neutral lipids. Biophys. J. **80**:1703-1711.

126. **Klasczyk, B., S. Panzner, R. Lipowsky, and V. Knecht**. 2010. Fusion-relevant changes in lipid shape of hydrated cholesterol hemisuccinate induced by pH and counterion species. J Phys. Chem. B **114**:14941-14946.

127. **Siepi, E.** 2010. Mechanism of Amphoteric Liposomes & Application for siRNA Delivery. Doctoral Thesis Naturwissenschaftliche Fakultät I - Biowissenschaften - Martin-Luther-Universität, Halle-Wittenberg.

128. **Panzner, S., S. Lutz, E. Siepi, and C. Müller**. 2008. Improvements in or relating to amphoteric liposomes. Halle/S., DE patent WO/2008/043575.

129. **Siepi, E., S. Lutz, S. Meyer, and S. Panzner**. 2011. An Ion Switch Regulates Fusion of Charged Membranes. Biophysical Journal **100**:2412-2421.

130. **Burnett, J. R. and P. H. Barrett**. 2002. Apolipoprotein B metabolism: tracer kinetics, models, and metabolic studies. Crit Rev. Clin. Lab Sci. **39**:89-137.

References

131. **Grundy, S. M.** 1998. Hypertriglyceridemia, atherogenic dyslipidemia, and the metabolic syndrome. Am. J. Cardiol. **81**:18B-25B.

132. **Zambon, A., B. G. Brown, S. S. Deeb, and J. D. Brunzell**. 2006. Genetics of apolipoprotein B and apolipoprotein AI and premature coronary artery disease. J. Intern. Med. **259**:473-480.

133. **Soutschek, J., A. Akinc, B. Bramlage, K. Charisse, R. Constien, M. Donoghue, S. Elbashir, A. Geick, P. Hadwiger, J. Harborth, M. John, V. Kesavan, G. Lavine, R. K. Pandey, T. Racie, K. G. Rajeev, I. Rohl, I. Toudjarska, G. Wang, S. Wuschko, D. Bumcrot, V. Koteliansky, S. Limmer, M. Manoharan, and H. P. Vornlocher**. 2004. Therapeutic silencing of an endogenous gene by systemic administration of modified siRNAs. Nature **432**:173-178.

134. **Crooke, R. M., M. J. Graham, K. M. Lemonidis, C. P. Whipple, S. Koo, and R. J. Perera**. 2005. An apolipoprotein B antisense oligonucleotide lowers LDL cholesterol in hyperlipidemic mice without causing hepatic steatosis. J. Lipid Res. **46**:872-884.

135. **Zimmermann, T. S., A. C. Lee, A. Akinc, B. Bramlage, D. Bumcrot, M. N. Fedoruk, J. Harborth, J. A. Heyes, L. B. Jeffs, M. John, A. D. Judge, K. Lam, K. McClintock, L. V. Nechev, L. R. Palmer, T. Racie, I. Rohl, S. Seiffert, S. Shanmugam, V. Sood, J. Soutschek, I. Toudjarska, A. J. Wheat, E. Yaworski, W. Zedalis, V. Koteliansky, M. Manoharan, H. P. Vornlocher, and I. MacLachlan**. 2006. RNAi-mediated gene silencing in non-human primates. Nature **441**:111-114.

136. **Rozema, D. B., D. L. Lewis, D. H. Wakefield, S. C. Wong, J. J. Klein, P. L. Roesch, S. L. Bertin, T. W. Reppen, Q. Chu, A. V. Blokhin, J. E. Hagstrom, and J. A. Wolff**. 2007. Dynamic PolyConjugates for targeted in vivo delivery of siRNA to hepatocytes. Proc. Natl. Acad. Sci. U. S. A **104**:12982-12987.

137. **Baigude, H., J. McCarroll, C. S. Yang, P. M. Swain, and T. M. Rana**. 2007. Design and creation of new nanomaterials for therapeutic RNAi. ACS Chem. Biol. **2**:237-241.

138. **Nishina, K., T. Unno, Y. Uno, T. Kubodera, T. Kanouchi, H. Mizusawa, and T. Yokota**. 2008. Efficient in vivo delivery of siRNA to the liver by conjugation of alpha-tocopherol. Mol. Ther. **16**:734-740.

139. **Swayze, E. E., A. M. Siwkowski, E. V. Wancewicz, M. T. Migawa, T. K. Wyrzykiewicz, G. Hung, B. P. Monia, and C. F. Bennett**. 2007. Antisense oligonucleotides containing locked nucleic acid improve potency but cause significant hepatotoxicity in animals. Nucleic Acids Res. **35**:687-700.

140. **Haupenthal, J., C. Baehr, S. Kiermayer, S. Zeuzem, and A. Piiper**. 2006. Inhibition of RNAse A family enzymes prevents degradation and loss of silencing activity of siRNAs in serum. Biochem. Pharmacol. **71**:702-710.

141. **Van Veldhoven, P. P. and G. P. Mannaerts**. 1987. Inorganic and organic phosphate measurements in the nanomolar range. Anal. Biochem. **161**:45-48.

142. **Seglen, P. O.** 1976. Preparation of isolated rat liver cells. Methods Cell Biol. **13**:29-83.

143. **Aurich, H., M. Sgodda, P. Kaltwasser, M. Vetter, A. Weise, T. Liehr, M. Brulport, J. G. Hengstler, M. M. Dollinger, W. E. Fleig, and B. Christ**. 2009. Hepatocyte differentiation of mesenchymal stem cells from human adipose tissue in vitro promotes hepatic integration in vivo. Gut **58**:570-581.

144. **Deissler, V., R. Ruger, W. Frank, A. Fahr, W. A. Kaiser, and I. Hilger**. 2008. Fluorescent liposomes as contrast agents for in vivo optical imaging of edemas in mice. Small **4**:1240-1246.

145. **Saad, M., O. B. Garbuzenko, E. Ber, P. Chandna, J. J. Khandare, V. P. Pozharov, and T. Minko**. 2008. Receptor targeted polymers, dendrimers, liposomes: Which nanocarrier is the most efficient for tumor-specific treatment and imaging? Journal of Controlled Release **130**:107-114.

146. **Chonn, A., S. C. Semple, and P. R. Cullis**. 1992. Association of blood proteins with large unilamellar liposomes in vivo. Relation to circulation lifetimes. J. Biol. Chem. **267**:18759-18765.

147. **Slepushkin, V. A., S. Simoes, P. Dazin, M. S. Newman, L. S. Guo, M. C. Pedroso de Lima, and N. Duzgunes**. 1997. Sterically stabilized pH-sensitive liposomes. Intracellular delivery of aqueous contents and prolonged circulation in vivo. J. Biol. Chem. **272**:2382-2388.

148. **Geary, R. S., T. A. Watanabe, L. Truong, S. Freier, E. A. Lesnik, N. B. Sioufi, H. Sasmor, M. Manoharan, and A. A. Levin**. 2001. Pharmacokinetic properties of 2'-O-(2-methoxyethyl)-modified oligonucleotide analogs in rats. J. Pharmacol. Exp. Ther. **296**:890-897.

References

149. **Phillips, J. A., S. J. Craig, D. Bayley, R. A. Christian, R. Geary, and P. L. Nicklin.** 1997. Pharmacokinetics, metabolism, and elimination of a 20-mer phosphorothioate oligodeoxynucleotide (cgp 69846a) after intravenous and subcutaneous administration. Biochemical Pharmacology **54**:657-668.

150. **De Oliveira, M. C., V. Boutet, E. Fattal, D. Boquet, J. M. Grognet, P. Couvreur, and J. R. Deverre.** 2000. Improvement of in vivo stability of phosphodiester oligonucleotide using anionic liposomes in mice. Life Sci. **67**:1625-1637.

151. **Farquhar, M. G.** 2006. The glomerular basement membrane: not gone, just forgotten. J Clin Invest **116**:2090-2093.

152. **Matsuzaki, J., M. Kuwamura, R. Yamaji, H. Inui, and Y. Nakano.** 2001. Inflammatory Responses to Lipopolysaccharide Are Suppressed in 40% Energy-Restricted Mice. The Journal of Nutrition **131**:2139-2144.

153. **Monia, B. P., K. W. Dobie, S. M. Freier, I. J. Popoff, W. S. F. Wong, and J. G. Karras.** August 2008. Antisense modulation of p38 mitogen activated protein kinase expression. Carlsbad, CA, USA patent US 2008/0194503.

154. **Cao, W. H., Y. Xiong, Q. F. Collins, and H. Y. Liu.** 2007. p38 mitogen-activated protein kinase plays a critical role in the control of energy metabolism and development of cardiovascular diseases. Zhong. Nan. Da. Xue. Xue. Bao. Yi. Xue. Ban. **32**:1-14.

155. **Patel, D. J., J. B. Ma, Y. R. Yuan, K. Ye, Y. Pei, V. Kuryavyi, L. Malinina, G. Meister, and T. Tuschl.** 2006. Structural biology of RNA silencing and its functional implications. Cold Spring Harb. Symp. Quant. Biol. **71**:81-93.

156. **Kunze, H., B. Hesse, and E. Bohn.** 1982. Effects of antimalarial drugs on several rat-liver lysosomal enzymes involved in phosphatidylethanolamine catabolism. Biochim. Biophys. Acta **713**:112-117.

157. **Zuidam, N. J., H. K. Gouw, Y. Barenholz, and D. J. Crommelin.** 1995. Physical (in) stability of liposomes upon chemical hydrolysis: the role of lysophospholipids and fatty acids. Biochim. Biophys. Acta **1240**:101-110.

158. **Dijkstra, J., M. Van Galen, and G. L. Scherphof.** 1984. Effects of ammonium chloride and chloroquine on endocytic uptake of liposomes by Kupffer cells in vitro. Biochimica et Biophysica Acta (BBA) - Molecular Cell Research **804**:58-67.

159. **Wu-Pong, S., T. L. Weiss, and C. A. Hunt.** 1992. Antisense c-myc oligodeoxyribonucleotide cellular uptake. Pharm. Res. **9**:1010-1017.

160. **Loke, S. L., C. A. Stein, X. H. Zhang, K. Mori, M. Nakanishi, C. Subasinghe, J. S. Cohen, and L. M. Neckers.** 1989. Characterization of oligonucleotide transport into living cells. Proc. Natl. Acad. Sci. U. S. A **86**:3474-3478.

161. **Jarve, A., J. Muller, I. H. Kim, K. Rohr, C. MacLean, G. Fricker, U. Massing, F. Eberle, A. Dalpke, R. Fischer, M. F. Trendelenburg, and M. Helm.** 2007. Surveillance of siRNA integrity by FRET imaging. Nucleic Acids Res. **35**:e124.

162. **Holen, T., M. Amarzguioui, E. Babaie, and H. Prydz.** 2003. Similar behaviour of single strand and double strand siRNAs suggests they act through a common RNAi pathway. Nucl. Acids Res. **31**:2401-2407.

163. **Rejman, J., V. Oberle, I. S. Zuhorn, and D. Hoekstra.** 2004. Size-dependent internalization of particles via the pathways of clathrin- and caveolae-mediated endocytosis. Biochem. J. **377**:159-169.

164. **Hillaireau, H. and P. Couvreur.** 2009. Nanocarriers' entry into the cell: relevance to drug delivery. Cellular and Molecular Life Sciences **66**:2873-2896.

165. **Roerdink, F. H., J. Regts, T. Handel, S. M. Sullivan, J. D. Baldeschwieler, and G. L. Scherphof.** 1989. Effect of cholesterol on the uptake and intracellular degradation of liposomes by liver and spleen; a combined biochemical and [gamma]-ray perturbed angular correlation study. Biochimica et Biophysica Acta (BBA) - Biomembranes **980**:234-240.

166. **Jaureguiberry, M. S., M. A. Tricerri, S. A. Sanchez, H. A. Garda, G. S. Finarelli, M. C. Gonzalez, and O. J. Rimoldi.** 2010. Membrane organization and regulation of cellular cholesterol homeostasis. J. Membr. Biol. **234**:183-194.

167. **Huong, T. M., H. Harashima, and H. Kiwada**. 1998. Complement dependent and independent liposome uptake by peritoneal macrophages: cholesterol content dependency. Biol. Pharm. Bull. **21**:969-973.

168. **Sanchez, S. A., M. A. Tricerri, and E. Gratton**. 2007. Interaction of high density lipoprotein particles with membranes containing cholesterol. J. Lipid Res. **48**:1689-1700.

169. **Bisgaier, C. L., M. V. Siebenkas, and K. J. Williams**. 1989. Effects of apolipoproteins A-IV and A-I on the uptake of phospholipid liposomes by hepatocytes. J. Biol. Chem. **264**:862-866.

170. **Yan, X., F. Kuipers, L. M. Havekes, R. Havinga, B. Dontje, K. Poelstra, G. L. Scherphof, and J. A. A. M. Kamps**. 2005. The role of apolipoprotein E in the elimination of liposomes from blood by hepatocytes in the mouse. Biochemical and Biophysical Research Communications **328**:57-62.

171. **Akinc, A., W. Querbes, S. De, J. Qin, M. Frank-Kamenetsky, K. N. Jayaprakash, M. Jayaraman, K. G. Rajeev, W. L. Cantley, J. R. Dorkin, J. S. Butler, L. Qin, T. Racie, A. Sprague, E. Fava, A. Zeigerer, M. J. Hope, M. Zerial, D. W. Sah, K. Fitzgerald, M. A. Tracy, M. Manoharan, V. Koteliansky, A. d. Fougerolles, and M. A. Maier**. 2010. Targeted Delivery of RNAi Therapeutics With Endogenous and Exogenous Ligand-Based Mechanisms. Mol Ther .

172. **Mahley, R. W. and Z. S. Ji**. 1999. Remnant lipoprotein metabolism: key pathways involving cell-surface heparan sulfate proteoglycans and apolipoprotein E. J. Lipid Res. **40**:1-16.

173. **Beisiegel, U., W. Weber, G. Ihrke, J. Herz, and K. K. Stanley**. 1989. The LDL-receptor-related protein, LRP, is an apolipoprotein E-binding protein. Nature **341**:162-164.

174. **Shimano, H., Y. Namba, J. Ohsuga, M. Kawamura, K. Yamamoto, M. Shimada, T. Gotoda, K. Harada, Y. Yazaki, and N. Yamada**. 1994. Secretion-recapture process of apolipoprotein E in hepatic uptake of chylomicron remnants in transgenic mice. J. Clin. Invest **93**:2215-2223.

175. **Hu, L., C. C. van der Hoogt, S. M. S. Espirito Santo, R. Out, K. E. Kypreos, B. J. M. van Vlijmen, T. J. C. Van Berkel, J. A. Romijn, L. M. Havekes, K. W. van Dijk, and P. C. N. Rensen**. 2008. The hepatic uptake of VLDL in lrp$^{-/-}$ldlr$^{-/-}$vldlr$^{-/-}$ mice is regulated by LPL activity and involves proteoglycans and SR-BI. J. Lipid Res. **49**:1553-1561.

176. **Forte, T. M., J. J. Bell-Quint, and F. Cheng**. 1981. Lipoproteins of fetal and newborn calves and adult steer: a study of developmental changes. Lipids **16**:240-245.

177. **Nishina, P. M., J. Wang, W. Toyofuku, F. A. Kuypers, B. Y. Ishida, and B. Paigen**. 1993. Atherosclerosis and plasma and liver lipids in nine inbred strains of mice. Lipids **28**:599-605.

178. **Tsutsumi, K., A. Hagi, and Y. Inoue**. 2001. The relationship between plasma high density lipoprotein cholesterol levels and cholesteryl ester transfer protein activity in six species of healthy experimental animals. Biol. Pharm. Bull. **24**:579-581.

Novosom internal reports **nov-025-2005, -012-2006, 001-2009 and 006-2009** are designated as confidential but limited access can be warranted upon request.

7. Abbreviations

Ab	antibody
Abb.	abbreviation
alc.	alcohol
ALT	alanine aminotransferase (ALAT)
apoB100	apolipoprotein B100
aqua dest	distilled water
ASO	antisense oligonucleotides
AST	aspartate aminotransferase (ASAT)
ATP	adenosine triphosphate
AUC	area under the concentration curve
Ave.	average
BD	biodistribution
BW	body weight
CA	citric acid
chol	cholesterol
CLSM	confocal laser scanning microscopy
CL_{tot}	total body clearance
C_{max}	maximal concentration
CMV	cytomegalovirus
conc.	concentration
d	days
DMGS	Dimyristoyl-glycero-succinic acid
dsRNA	double-stranded RNA
EDTA	ethylenediaminetetraacetic acid
EFM	epifluorescence microscopy
ELISA	enzyme-linked immunosorbent assay
EtOH	ethanol
EU	endotoxin units
FDA	US Food and Drug Administration
FRET	fluorescence resonance energy transfer
HDL	high density lipoproteins
IFNγ	interferon gamma
IL-1ß	interleukin-1 beta
IL-6	interleukin-6
inj.	injection
iv	intravenous

KD	knockdown
kDa	kilo-Dalton
LDH	lactate dehydrogenase
LDL	low density lipoproteins
LNA	locked nucleic acid
LNP	lipid nanoparticle
mer	from *meros* [greek] = part; meaning the length of an oligonucleotide
miRNA	microRNAs
m/f	male & female
mg/kg	milligram substance per kilogram (mg/kg)
MPS	mononuclear phagocyte system
MOE	methoxy-ethyl modification at 2' position of the ribose (2'MOE)
MW	molecular weight
NIR	near-infrared red
OD	optical density
oligo	oligonucleotide
o/n	over night
PAA	polyacrylamide
PAGE	polyacrylamide gel electrophorese
PEG	polyethyleneglycol
PES	polyethersulfone
PK	pharmacokinetic
PPIB	peptidylprolyl isomerase B (cyclophilin B)
PMH	primary mouse hepatocytes
PS	phosphorothiolated DNA backbone
P/S	penicillin & streptomycin
QG	Quantigene mRNA analysis
qPCR	quantitative (real time) PCR
RT-PCR	reverse transcription polymerase chain reaction
scr	scrambled (mismatch) control siRNA
SDS	sodium dodecylsulfate
SEM	standard error mean
siRNA	small interfering RNAs
$t_{1/2}$	half-life
TNFα	tumor necrosis factor
w/	with
WB	Western Blot analysis
w/o	without

8. Appendix

8.1 Renal uptake of free, non-encapsulated Cy5.5-labeled ASO

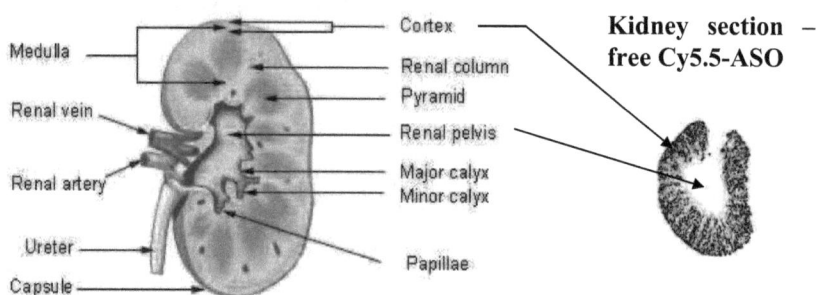

Kidney section – free Cy5.5-ASO

Fig. 8.1: Schematic presentation (left) of a frontal kidney section and NIR-scan of a croysected kidney treated with free Cy5.5-labeled ASO (right). Cy5.5-labeled ASO predominantly distributes into the kidney cortex whereas no signal is visible in the renal pelvis. Schematic kidney section taken from: www.wikidoc.org/index.php/Image:Illu_kidney2.jpg

8.2 Lipids for rational design

All lipids used in this work are synthetic, HPLC purified and solvent free substances and were provided as dry powder except for DOGS which has an oily appearance. Lipids were purchased from the following manufacturers: DC-Chol from Sigma Aldrich (Saint Louis, MO, USA); from Lipoid (Ludwigshafen, Germany); DOTAP from Merck Eprova AG (Schaffhausen, Switzerland); DOGS from Avanti Polar lipids (Alabaster, AL, USA); CHIM, Chol-C4N-Mo2 and HisChol from Chiroblock GmbH (Wolfen, Germany).

Abb.	MW	Full name
	pK	
Tail vol. [Å^3]	Head vol. [Å^3]	Chemical structure
CHIM	537.8	Cholesterol-(3-imidazol-1-yl propyl)-carbamate
	6.50	
334.0	119.2	

Appendix

Chol-C4N-Mo2	613.9	[(2-Morpholin-4-yl-ethylcarbamoyl)-ethyl]-carbamic acid cholesterylester	
	6.53		
334.0	195.3		
DC-Chol	537.3	Cholesteryl 3β-(N-[dimethylaminoethyl] carbamate) (chloride)	
	7.56		
334.0	87.2		
DOGS	721.1	1,2-Dioleoyl-sn-glycero-3-succinate	
	5.33		
511.8	90.2		
DOTAP	698.5	1,2-Dioleoyl-3-trimethyl-ammonium-propane (chloride salt)	
	15.00		
511.8	57.2		
HisChol	579.9	(α-(3'O-cholesteryloxycarbonyl)-δ-(4- ethylimidazole)-succineamide)	
	6.67		
334.0	150.5		

Tab. 8.1: Lipid abbreviations & full name, structures, partial molecular volumes and pK values.
Lipid synthesis details are available upon request.

8.3 Lipid composition and parameters of cellular transfectants for primary mouse hepatocytes

nov#	Amphoter Class	Lipid composition	mol%	κ_{min}	IC_{50}
-	I	DODAP:DOGS:Chol	24:36:40	0,132	25
-	I	DODAP:DMGS:Chol	20:30:50	0,137	80
-	I	DODAP:Chems:Chol	20:30:50	0,141	350
-	I	DOTAP:DOGS:Chol	15:45:40	0,141	11
nov729	I	DODAP:DMGS:Chol	24:36:40	0,145	22
-	I	DOTAP:Chems:Chol	31:39:30	0,151	30
-	I	DODAP:DMGS:Chol	15:45:40	0,152	300
-	I	DODAP:Chems:Chol	25:45:30	0,153	300
-	I	DOTAP:DMGS:Chol	15:45:40	0,156	600
-	I	DODAP:DMGS:Chol	36:54:10	0,164	40
-	I	DOTAP:DMGS:Chol	18:52:30	0,165	60
-	II	CHIM:DMGS:Chol	20:20:60	0,165	800
-	I	DOPE:DODAP:DMGS:Chol	13:24:36:27	0,169	100
-	II	MoChol:DMGS:Chol	20:20:60	0,184	≥1000
-	II	HisChol:DOGS:Chol	20:20:60	0,185	400
-	I	DC-Chol:DMGS:Chol	26:39:35	0,188	≥1000
-	II	HisChol:DMGS:Chol	20:20:60	0,195	400
-	II	HisChol:DMGS:Chol	15:45:40	0,205	≥1000
-	I	DC-Chol:DMGS:Chol	36:44:20	0,211	≥1000
-	II	HisChol:DMGS:Chol	30:20:50	0,228	600
-	II	CholC4N-Mo2:DMGS:Chol	23:47:30	0,239	≥1000
-	II	POPC:DOPE:HisChol:DMGS:Chol	7:28:25:30:10	0,300	≥1000
nov038	II	POPC:DOPE:MoChol:Chems	15:45:20:20	0,350	≥1000
-	II	POPC:DOPE:MoChol:Chems	6:24:47:23	0,369	≥1000

Tab. 8.2: **Lipid composition and parameters of cellular transfectants for PMHs.** For gene silencing studies ApoB I 5'P siRNA was encapsulated into liposomes. After production, liposomes were dialyzed and concentrated using MicroKros hollow fiber membranes. IC_{50} values were calculated from quantified apoB100 mRNA levels of a distinct dose range (10...1000 nM).

8.4 Transfection inhibition by lipoproteins

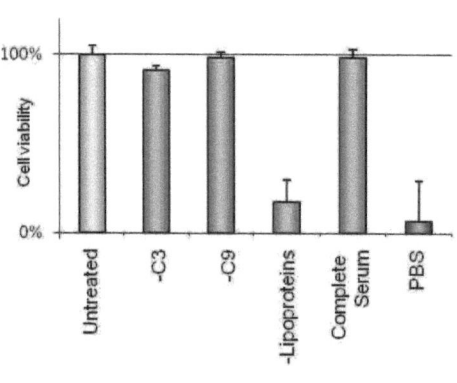

Fig. 8.2: **Transfection efficiency of nov729 is totally diminished in the presence of lipoproteins.** Full activity is restored upon depletion of lipoproteins from serum. Nov729 was pre-incubated with human serum devoid of complement factors or lipoproteins for 30 min. HeLa cells were treated with nov729 encapsulating PLK1 siRNA at a concentration of 50 nM. Cell viability was determined 72 h post-treatment. In the absence of serum (PBS) or lipoprotein-depleted serum nov729-Plk1 led to an inhibition of cell proliferation. (unpublished data, prepared by E. Siepi)

9. Acknowledgement

I owe my deepest gratitude to Dr. Steffen Panzner for giving me the opportunity to accomplish my PhD work at novosom. I gratefully thank you for the supervision, for keeping my thoughts focused and for scientific inspiration in the past years.

I am indebted to Prof. Dr. Sven-Erik Behrens for being the primary reviewer of this thesis, for providing relevant comments and enabling the success of the dissertation. Further, I owe Prof. Dr. Ingo Heilmann and Prof. Dr. Philipp Wiedemann my debt of gratitude for the assumption of the accessory expertise.

I also gratefully acknowledge the enlightening scientific and non-scientific discussions with Drs. Una Rauchhaus, Silke Lutz, Andreas Dieckmann, Ludger Ickenstein and Gerold Endert. The superb technical assistance of Ute Vinzenz, Claudia Müller, Katrin Blanke and Manuela Dammer is also acknowledged.

This work would not have been possible without the dedicated efforts of the staff and personnel at Preclinics GmbH with special thanks to Dr. Simone Odau and Jonas Füner. At this point I would like to thank the lab members from Probiodrug for providing technical equipment and Drs. Nadine Stöhr and Hendryk Aurich from ZAMED for her help and advice at the CLSM and isolation of mouse hepatocytes.

It is a pleasure and a deep desire to thank those who supported me in a number of ways: Krischi, Jan, Steffen H. & Holger and my parents whose time, interest, patience and belief I could call at any phase of the thesis.

My beloved children, Clara und Jonas, I want to thank for their unconditioned love, balmy smiles and cheerful welcome at the end of a taxing working day. You led me through the winter time of this project.

In genuine gratitude and admiration I will dedicate the final lines to you, Juliane. We succeeded a long and stony path together for many years, with the achievement of bringing both, our children and our dissertations to the world. Our persistence is always the way for us to overcome difficulties.

"I can be an a..hole of the grandest kind, I can withhold like it's going out of style! (...)
I blame everyone else, not my own partaking. My passive-aggressiveness can be devastating.
I'm terrified and mistrusting, and you've never met anyone as, as closed down as I am sometimes.
You see everything, you see every part. You see all my light and you love my dark!
You dig everything of which I'm ashamed, there's not anything to which you can't relate!
And you're still here!"
(Alanis Morissette – "Everything")

i want morebooks!

Buy your books fast and straightforward online - at one of world's fastest growing online book stores! Environmentally sound due to Print-on-Demand technologies.

Buy your books online at
www.get-morebooks.com

Kaufen Sie Ihre Bücher schnell und unkompliziert online – auf einer der am schnellsten wachsenden Buchhandelsplattformen weltweit! Dank Print-On-Demand umwelt- und ressourcenschonend produziert.

Bücher schneller online kaufen
www.morebooks.de

VDM Verlagsservicegesellschaft mbH
Heinrich-Böcking-Str. 6-8   Telefon: +49 681 3720 174   info@vdm-vsg.de
D - 66121 Saarbrücken   Telefax: +49 681 3720 1749 www.vdm-vsg.de

Printed by Books on Demand GmbH, Norderstedt / Germany